To - Joyce Weimer   9-25-2013

Dick's Book

~

Reflections of a Farmer

Samuel Dick Harris

By
Samuel Dick Harris

Foreword by
Samuel D. Harris, Jr.

Dementi Milestone Publishing

DEDICATION

This book is dedicated to my wife, Shirley, who, for the last forty-five years of my life, has provided for all my needs for a good and happy life and whose trust and love I value very much.

Dick's Book

1st Printing

© copyright 2013

Written by Samuel Dick Harris
Mineral, Virginia

Cataloging-in-publication data for this book is available from the Library of Congress
ISBN: 978-0-9889099-9-1

All rights reserved. No part of this book may be reproduced or transmitted in any form or by any other means, electronically or mechanical, including photocopying, recording, or by any information storage and retrieval system, without the written permission of the Publisher.

To order a copy, or for information, please visit:
www.dementibooks.com

Cover design and manuscript design by Dianne Dementi

Printed in the USA

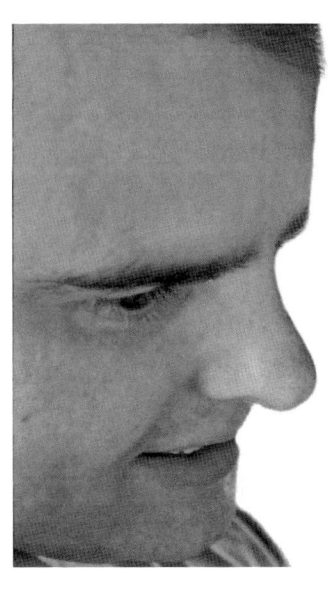

Foreword

My Father, Samuel Dick Harris, was born on March 3, 1932 in the midst of one of the most pivotal points in our nation's history. America was on the verge of beginning a long recovery out of the Great Depression. In a short nine years, the country would be entering the greatest conflict the world has ever known, World War II. It was a time of conflict and uncertainty, but a time of great change and innovation as well. Times were tough, yet, as always, the American people persevered.

This book represents his memories of growing up on a small dairy farm in Spotslyvania, Virginia, in the 1930s through the 1950s. While reading his manuscript for the first time, I was struck by the amount of information he recalled from life in general, but also by the very precise detail he included on "how things worked and how they were designed." He explains in great detail how the advent of electricity and advancements in farming technology transformed farm life. These advancements allowed farmers to produce more crops and cows to produce more milk than ever before. Unfortunately, the age old laws of "supply and demand" have created a situation where farmers have become victims of their own success. This book details his memories and observations of life on the farm back then as compared to how things are today, coupled with his advice for future generations. Always more curious than a cat, with an affinity for machinery and how things work, he paints vivid pictures of his early life. These experiences would go on to shape his life and career as an innovator in farm machinery in his own right. After reading his memoirs, I'm proud and honored to call him "Dad" and to pass on his story to future generations.

Samuel D. Harris, Jr.

ACKNOWLEDGMENTS

  I would like to thank Barbara Sprouse for typing the first draft of this book from my cursive writing, my son Samuel and his helpmate Jann Edmondson for putting the original typed draft into a computer format, David Black for a critical review and many suggestions for improvement, Steve Weeber for Wetach pictures, Sam Moore for a picture, Francis Baker for Holladay Mill pictures, Charles Harris for the farm aerial image on the back cover, Step-Daughter Judy Stephenson for revisions with the computer, Chris Owens and Bruce Stone for their review and opinions, and, last, Wayne Dementi for picture work and guidance. Without their help, this book would never have happened, and I am deeply grateful for their help.

PREFACE

If you wonder what this book is about, it is about country and farm life in central Virginia in the 1930's and 1940's. A transition time of how life and work was done then. I share how farming with animal power and home life without electricity converted or transitioned to life with new forms of power to do manual work. Also, I review how electricity changed and improved home life.

It is about getting the first farm tractor and related farm machined for the farm to the getting of electricity and indoor plumbing for the house. It was about how life was then.

It is about how crops - corn, hay, wheat and oats – were planted, cultivated and harvested with most tasks done by hand.

It is about milking cows by hand and building a state-of-the-art dairy and getting milking machines.

It is about gardens and chickens and turkeys and my mother running a household on egg money. It is about $1.00 a day wages for labor plus a noon meal. We called it dinner.

It is about World War II and the sacrifices people had to make because of the war and what "Army Flash" meant when spoken into the party line telephone. It is about dirt roads and building new roads with bulldozers. It is about cutting trees without chainsaws and how log carts and sawmills worked. It is about killing hogs in the winter and kids going barefoot in the summer months.

Sprinkled throughout this book are references to God and my belief in him. At the end, I share a testimony told to me by a customer (now deceased), who had an out-of-body experience. I also explain how I have come to believe that the Book of Revelations is the Apostle John's out-of-body experience.

When I set out to write this book in the spring of 2007, having idle time on my hands from open heart surgery, I wanted to make a record of what country and farm life was like in central Virginia as I remembered it. It covers the years 1936 to 1956. I always kept in mind that the book would be read by young people and urban people who do not have a clue about how farm life was before electricity, tractors, farm machines or paved roads. This is the reason some sections of the book are in detail and may be boring to some readers and yet fascinating to others. I wanted to explain to young people how things were done back then. I hope it will bring back memories to those people who lived during that time and experienced what the book is about.

The book had no title in the beginning. In the first attempt to make it a book, by my son Samuel, he named it "The Memories of a Retired Dairy Farmer." This might make a subtitle and is partly correct, but it is way too long. I wanted a short, simple title and could not come up with anything I liked. The my step-daughter, Judy Stephenson, got involved making corrections and revisions on the computer. Judy is very good with computers and can make it do what it was designed to do. I did not trust myself to make changes for fear of hitting the wrong key and wiping out a lot of work. In doing her work, she always referred to the manuscript as "Dick's Book." After hearing it being called by that name a few times, I began to like it and so its name. I hope it is an accurate record so that those who come after us will have some idea of how life was when electricity first came into use and tractors replaced horses for farm work.

Samuel Dick Harris

Table of Contents

Chapter One ~ Tractors and the Demise of the Farm Family 1

Chapter Two - The House 6

Chapter Three ~ Army Flash 15

Chapter Four ~ Radio 20

Chapter Five ~ Barefoot 23

Chapter Six ~ Hog Killing 25

Chapter Seven ~ Food Refrigeration 30

Chapter Eight ~ 1940 Snow 33

Chapter Nine ~ Garden and Mill 36

Chapter Ten ~ School 42

Chapter Eleven ~ Dairy 46

Chapter Twelve ~ Corn 68

Chapter Thirteen ~ Seed Corn 77

Chapter Fourteen ~ Silage 79

Chapter Fifteen ~ Hay 84

Chapter Sixteen ~ Wheat and Oats 97

Chapter Seventeen ~ Chickens and Turkeys 106

Chapter Eighteen~The Sawmill, Logging and the Forests 112

Chapter Nineteen ~ New Roads and Bridges 122

Chapter Twenty ~ Pedro 133

Chapter Twenty-One ~ Grandpa Abner 136

Chapter Twenty-Two ~ Grandpa Dick 140

Chapter Twenty-Three ~ A Testimony 144

Chapter One

Tractors and the Demise of the Farm Family

I did not realize it at the time, but I came into the farming scene in the mid-stages of the transition from the widespread use of animals (horses and mules) to perform field work to the use of petroleum-powered machines. One example of this transition was the switch from the slow, hand-worked method of growing corn to performing the same tasks with machines where the product is never touched by hand. Corn is just one of the farm products that have been profoundly affected by mechanization. Almost every aspect of country life has changed in the last one hundred and fifty years, a relatively short span in the scheme of history. These changes have almost seemed to take place overnight. In virtually every facet of farm life that I describe in this book, be it acquiring electricity or the dairy and all the crops, the use of mechanical power has allowed a person to produce many times more product than could be produced through hand or animal power. This ability to produce a product from the earth with very little human effort compared to with hand tools has profoundly changed farm life.

Over one hundred years ago in 1907, the word "tractor" was first used. The Hart-Parr Company in Charles City, Iowa, was the first company to refer to these machines as tractors. They were very slow, crude, cumbersome machines, but they did function and they did replace animals as a source of power. For many years, the tongues of horse-drawn machines had to be shortened and converted

John Deere tractors 1956 - one Model B, one Model GM and two Model 50s

to connect to a tractor drawbar for pulling. Another early use of tractors was the belt-pulley systems where a long, flat belt would be used to transfer power from the tractor to various other farm machines. Machines such as threshing machines, sawmills, feed grinders, etc. were stationary machines that required rotary power. Prior to tractors, large steam-powered machines were used for this purpose. These steam-powered machines were slower and even more cumbersome than early tractors. With the exception of powering a sawmill, I never saw one of these steam engines actually perform farm work. Today, you can see them in action at antique farm shows across our land, typically during the summer months. It was not until the 1930's that Harry Ferguson, an Irish engineer and inventor, got to thinking "Why not mount the machines?" that they were pulling with a tongue directly onto the back of the tractor. He presented his system to Henry Ford in 1939, and they struck a deal for Ford to build tractors with a three-point hitch system. Eventually, every tractor manufacturer adopted some form of Ferguson's ideas.

  My oldest brother, W.O., purchased a John Deere "B" tractor in late 1940. Daddy was always a horse person; he never learned to operate a vehicle. When he was growing up, he learned about horses at a young age, as that was all they had for transportation and work then. When you can only plow one acre a day with a team of horses and then take a tractor with a two-bottom plow and plow ten acres in the same amount of time, it doesn't take a young person long to switch. W.O. was young and open to change; Daddy, on the other hand, stuck with what he knew. You see the same type of thing today with computers: most of my generation do not know how to effectively operate one, whereas a six-year-old does. The tools that W.O. eventually purchased for the tractor were a two-bottom plow, a disc, a two-row corn cultivator, and a seven-foot hay mower. The tractor and the plow were purchased first. This was how we made the transition from horses and other farm animals to power machinery with tractors.

When W.O. was researching the purchase of a tractor, he had the Harlows from Mineral bring their "B" tractor to our farm for a demonstration. In addition, he had Mr. Sam Hudson from Orange bring a Farmall "H" tractor for demonstration purposes, as well. I didn't get to observe the demonstration myself as I was in school, so I'm not sure why he selected the John Deere "B." From then on, nearly every farm machine we purchased was a John Deere. I once heard a person remark that if W.O. had been given a tractor that commanded this brand loyalty, that company would have been many thousands of dollars ahead over all the years he purchased farm machinery. Thus, the gift would have a paid for itself many times over due to the future sales it would have generated. In a strange twist of fate, the very same "B" tractor W.O. purchased in 1940 was purchased for me by my wife Shirley from Mr. J.D. "Mac" McClary in the 1970's for roughly $250. I still own it to this day.

In late 1945, we purchased a John Deere "GM" tractor from the Harlows. The "GM" was nearly twice the size of the "B". We then bought a John Deere chopper in 1946. Sometime prior to 1945, we bought a John Deere 12A combine, as well. The purchase of farm machines never seemed to end. There was always a bigger or better one to be had. This is still happening today, but not as rapidly as it was fifty years ago. What has happened with farming is that it is nearly impossible to start from zero as we did in 1940 and gradually grow into a business of half a million in yearly gross. If a young couple purchases bare property like my father did and has to construct a house, barns, and even buy used machinery to work with, what they could produce with their efforts would have such a small profit margin, they would go broke. With all of the struggle and sacrifice that would be required, they would have a very difficult time in making a better life for themselves. Comparing the prices that farm products brought sixty years ago with present day prices for the same product will show that they have nowhere near kept up with the increase in costs. The economic law of supply and demand states that the more of a product you have, without a corresponding increase in demand, the less the product is typically worth. Through the widespread use of machines, a virtual flood of farm products is produced. With these large quantities of goods and no corresponding increase in demand, prices fall. This simple rule will always hold true.

When I leave a farm trade show which I sometimes attend, I come away thinking that, everything a farmer needs is on display except the one thing he needs most. What the modernday farmer needs most is something that would allow the price of what he produces to keep up with the price of all the new machines that he has just looked at. Nobody has the answer to this dilemma. The various government programs are supposed to do this, but in reality none of them has worked for the long term. Things have unintended consequences, which with time show up and usually are not good. The ultimate judge of any economic plan is the market for whatever commodity they are designed to help. It would take an entire book about economics to answer this question. I do not expect to take on this subject in this writing. The person who could answer the dilemma the modern farmer has would have to be very wise. For myself, I have become aware of the transition I have witnessed in agriculture.

What is the cause of the demise of the farm family? Was it caused by the invention of various mechanical machines that allowed a person to produce so much more with the same amount of work that I witnessed growing up? Do we want to do the impossible and go back to the days of the horse and pitchfork?

In my opinion, the farm is the best place to raise children. Children can learn more about so many different facets of life when growing up on a farm. They can learn how to work, and to be responsible, and they can acquire these skills far sooner than children raised in an urban environment. It is a known fact that young men that come from the country with a farm background make the best workers. This pool of country boys has disappeared now that most people do not grow up on farms any longer.

What lies ahead for people who want to live and raise their families on the farm? What will our country be like due to the great shift in population from agriculture to urban/suburban living? Do I attempt to answer such a profound question?

For starters, I think that the transition of people leaving the farm has ended. Most of the rural population now is not directly employed in agriculture, although they may work for a business that depends on agriculture. All the machines used in farming require people to build them, maintain them, and fix them when they stop running. Another trend I see is for people to retire to rural areas for the later years of their lives. I foresee the USA becoming more like the so-called third world countries that have a great divide between the extremely wealthy and the average citizen. I've read studies that have shown that middle-class income families as a percentage of the entire population is on the decline in the US. Percentage-wise the number of families that can afford, homes worth $500K and up is on the decline. There will always be a few that can afford such, but those numbers are on the decline. If one is so lucky as to be born into such wealth, they will have to be on guard their entire life so as to not squander it. It has been said, "It's not so much as what you have, it's what you do with what you have that counts."

There are certain niches in which one can make a life in the country. Some people that seek or want a country life will discover what their talent is and find their niche. I don't foresee the number being anywhere remotely near the rural population numbers of sixty to seventy years ago. There's no going back to the good old days. "The good old days" for every generation is typically between the ages of ten to thirty. Everyone looks back to how life was like in his youth. This is why people collect antiques, attend antique shows, etc. because they serve as a vivid reminder that brings back memories acquired during their youth.

Chapter TWO

The House

My father built our house in 1915 on the land he bought from Cherry Grove. He kept a diary for the year and recorded all the steps of its construction. Two brothers, Mr. Will Holladay and Mr. Jim Holladay, were the main carpenters. Mr. Henry Williams served as the main worker and an assistant carpenter. Father would hire additional workers as needed to assist the three main builders. The house, which is still in use today, is a two-story frame house with a crawl space between the ground and the first floor. It was built with a front porch that ran most of the way across the front of the house. It has a screened-in back porch located on the south side of the rear of the house.

Farm House - Built in 1915. South side shows both front porch to right and back porch on left. 1940

The House

Front view of house. 1940

    The house was constructed in a T-shape with the front of the house being the top of the T. When originally built, it had eight main rooms, plus two stairs and two hallways. The roof was a standing-seam metal roof which is still in good service today (2007) and has required only painting over the years. One fond memory I have is when we had a steady rain, how the water coming off the roof hit the back porch roof right by my window and how it was easy to fall asleep to the sound of rain falling on the metal roof. You don't get this sound of rain falling on other types of roofs.

    The outside of the house was a weatherboard construction. The inside was lath and plaster walls with pine wood floors. It had no insulation anywhere. I don't think insulation existed in 1915. It was heated by fireplaces in the two front bottom-floor rooms. The rest of the rooms had thimbles in a chimney (it has three chimneys). A wood stove could be set in the room and a stove pipe used to connect the stove to the chimney.

    You opened the windows for cooling and hoped a breeze would blow in the summertime. In the summer, chimney swallows (a migratory bird) would build a nest in the chimney and raise a family. These birds can fly very fast, and I spent many summer evenings, at the edge of dark, watching for when these birds would come out and fly and dart very quickly about catching some kind of insect. Sometimes the birds would fall down the chimney into the fireplace and get inside the house. They would fly around in the room trying to get out. Sometimes opening a window would work. Other times, you would get a broom and pin the bird to the wall, then gently catch it in your hands and take it outside and release it. They left in the fall and came back after the winter was over.

In 1940, we had the opportunity to sign up for electricity. This was for Rural Electrification Administration service out of Bowling Green, Virginia. They routed and built the lines. The landowner gave them the right-of-way to let them run their lines across his property. If a person did not want electricity or did not want the lines run across his property, then they had to reroute the lines so as to miss his property. Sometimes a person would change his mind and let the lines be run for his neighbors. Each customer had to have his house wired at his expense. The power company would run its lines to the meter. The customer had to do everything beyond the meter. People didn't say they were going to get electricity but instead would say they were going to get "lights." The first primary use of electricity was for light, followed by power for the radio, a refrigerator and an electric iron to iron clothing. Now everything depends on electricity.

My father got Mr. Frazier Thompson and Mr. Curtis Dickenson to do the wiring and plumbing of the house. At the same time this wiring was done, it was decided to put in two indoor bathrooms and an electric pump for the well. Before these indoor bathrooms were built, life was very different. Prior to indoor plumbing, for a bath, water was heated on a stove and then poured into a wash tub. You mixed cold water with the hot water to get to the temperature you liked. A wash tub was a round metal vessel about thirty inches across and twelve inches deep, made out of galvanized thin metal. It was placed in the room where you were going to take your bath. You sat down in the tub and took a bath the best you could. The dirty bath water was poured out in the yard after you were finished. For the toilet, we had a small building behind the house. People called it the johnny house or outhouse. The Civilian Conservation Corporation built us a nice new one out of pine sometime in the 1930's. Any building built out of untreated pine and set on the ground does not last many years, maybe ten years at the most.

The modern day equivalent of a johnny house is a port-a-john. They are both about the same size, but the big difference is that where the port-a-john is pumped out time to time, the johnny house was set over a pit or a hole dug in the ground. This pit was four or five feet deep and a little smaller than the house itself so the house wouldn't fall into the hole, which would be a disaster. It had a toilet seat with a lid. They had a square wooden vent pipe that went up through the roof on its back side. When the weather was very cold and snowy, or someone was sick, you used a slop jar or pot in the house where it was warm. We have a crude expression, "They don't have a pot to pee in," meaning the person is very poor in things of this world. Both the pot and the slop jar had to be taken to the outdoor toilet and emptied daily and then washed out. This was a very unpleasant job, with the smells and all, but still it had to be done. Life is much more pleasant with indoor toilets and plumbing with hot and cold running water. Even today, a good many people on this earth do not have this luxury. We tend to forget this fact.

It was fortunate that the interior design of daddy's house had two areas, one on each floor, that were of nearly the right size for a bathroom. For the upstairs it was at the end of the large hallway, which was big enough to put the bathroom in, plus still have a walkway into a room we called the "spare room." It was a guest room. Downstairs they converted a room called the "little room" into a bathroom. I don't know what its purpose was when the house was originally built, but it made a perfect bathroom.

Because the house had a three-foot crawl space and an attic, and the walls had no insulation, the pipes and wires could be run from the bottom of the house to the top without too much difficulty. I don't remember any walls having to be torn except for small holes. The two people that installed the wiring and plumbing worked at it part-time, as they were full-time farmers. Keep in mind that the house was twenty-five years old when this was done. Farm work always took precedence over anything else. Some weeks they never even showed up. I think they took about one year to complete these two projects, the plumbing being the more difficult.

For the sewage pipes, they used cast-iron pipes and poured melted lead to seal the pipe joints. No plastic pipes existed then. For the hot and cold water lines they used mostly copper pipes and soldered the joints. Some of the large water line pipes were threaded, galvanized pipes. The pit for the septic tank and drain fields lines were dug by hand. In addition, there was some digging in the crawl space under the house, as well as ditches to the septic tank. All this hand digging was performed by Charlie Hix. Charlie Hix was an elderly black man, and all I remember him doing was some kind of digging. If anyone died in the community, the family would get Charlie Hix to dig the grave. When he wasn't digging some kind of construction project, he worked at digging stumps up in a land-clearing project. Backhoes did not exist then. Charlie Hix had a son named William who some people said was absolutely worthless. Charlie Hix seemed to work all the time, William never did anything. William's only redeeming quality was that he had the reputation of being good at handling a team of horses or mules. It's strange how genes sometimes don't get passed down to the next generation.

A jet pump with a water tank was put under the house in the corner next to the well. A jet pump works by having two pipes that go down into the well. The pump itself mounts somewhere near the well, but is on top of the ground, not down into the well. A foot valve is put on at the very bottom of the two pipes that go down into the well. One of these pipes sends water down into the well to a small jet located on top of the foot valve. As the water is forced at a rapid rate through this small jet, it picks up more water which came back up the other pipe and goes into the holding tank. A jet pump has to be primed by pouring water into it before it will pump. The pump was powered by an electric motor.

At this time we had a Home Comfort wood cook stove. To have hot water for this new water system, a small water tank was mounted on one side of the wood stove fire box. When a fire was built in the cook stove, which was most of the time, it would heat the water in this small tank. This hot water was piped to the top of a thirty to forty gallon water tank located very near the cook stove. This hot water going in the top and cold water coming out the bottom would gradually fill the entire tank with hot water. No pump was required, as it worked on the principle that hot water rises.

Mother got a kitchen sink so we didn't need to use dish pans any longer. When the water system was finally finished and we had electricity, we could take a bath in a big tub, big enough to stretch out in. The upstairs tub had a shower, if you liked showers. Man, indoor plumbing was quite a luxury! No going outside to the cold outhouse.

One thing that amazed me the most was the vast difference between the oil lamp and just one electric light bulb lamp. There was a huge difference in the amount of light you get. Lamps, had to be filled with oil, wicks had to be trimmed, and the globes had to be washed from time to time. There was always the danger of knocking one over and starting a fire. With this new plumbing system put in the house and the electricity for lights and power, life was much more enjoyable. Do we really appreciate them now? Let the electricity get knocked out by a storm for an extended period of time and you will get a small taste of what life was like without it.

Mother had a round-tub Maytag washing machine that was powered by a kick-starter two cycle gas motor that made a lot of smoke when it ran. In addition to the smoke it produced, it was noisy. After we got electricity, it was converted to an electric motor, which was much quieter. Mr. Duncan in Louisa was a Maytag dealer, and he serviced the Maytag washing machine. I remember he and Mom discussed what to do with the old gas engine. Mr. Duncan really didn't want it, and neither did my mother. He ended up taking it, and now I wish I had spoken up and asked for it, as they are valuable antiques now. Before Mother got the Maytag with the gas motor, she must have washed by hand with a scrub board, but I don't remember that. This Maytag machine had a wringer that could be turned to several positions around its stem and locked in the position you picked. The wringer could also be reversed by a little lever. Monday was washday unless it was rainy weather. We had a clothes line outside for drying clothes. It was no point in washing clothes if you couldn't dry them, so if the weather was so bad that you couldn't dry clothes outside on the line, the washing was put off to another day.

Farm House. Screened in back porch. 1936 picture showing 1935 Ford. Note well to left with chain and pulley to lift bucket.

The washing machine was kept on the back porch where the washing was done. The set-up was such that we had a bench that could hold two wash tubs (the tubs we used to use to bathe in). They were set up behind the washing machine. The rinse water was put in one tub, and the second tub was used to catch the washed clothes. The clothes to be washed were first sorted into three groups. The first group was all-white things like sheets, shirts, and things not very dirty. They were washed first in the clean water. The second group of clothes was colored materials and again things that were not very dirty. The third group consisted of the dirty overalls, coveralls, and coats used in working outside on the farm.

The machine had a dasher in the center of the round tub. This dasher turned back and forth, first a partial turn one way, then a partial turn the other way. It made these partial turns very quickly. It was controlled by a lever, and the dasher could be put in neutral so that it didn't turn at all. After a batch was put in the washer and run for ten to fifteen minutes, the dasher was stopped and the wringer was put in gear. Each piece was fed by hand into this wringer. The wringer had two rubber rollers about two inches in diameter that turned in opposite directions. You would put the edge of the garment up to the slot made by the two rubber rollers that was turning in opposite directions. These rollers would catch the material and feed it through, thus squeezing out the water.

One had to be careful to keep your hands out of these rollers, as they would pull your hand and your arm into them. They were on springs so they could open up if something big got between them, but they could still hurt you if you got caught. If you did get caught the rollers could be stopped and put in reverse, then you could back your hand or arm out, assuming you had the presence of mind to do this.

The squeezed clothes went over into the rinse-water tub. You would take your hands and push the load back and forth in the rinse tub, maybe turning them over. Then you would swing the wringer into position so that when the clothes from the rinse tub were fed through the wringer, they then fell into the catch tub or basket, which was used to take them outside to hang on the line to dry. What this process did was use mechanical power to swish the clothes back and forth to get the dirt out as well as provide a mechanical method to get most of the water out. Each step required some hand work. Today, clothes are loaded into the washer and the time set, they are then unloaded when the machine has finished, ready to hang on the line or put in a dryer. There's no dangerous handwork feeding a wringer.

Chapter Three

Army Flash

We first received telephone service sometime in 1940 just before we received electricity. A "party line" was run across the North Anna River from Louisa County. This line ran all the way to the Louisa Courthouse where it was hooked to a switchboard that was manned by operators (mostly women) twenty-four hours a day, seven days a week. There were about eight or ten households hooked onto this phone service for the party line we shared. Our phone number was 19J3. The line had a 19-J side and a 19-W side. I have read that if this country used this type of service today, it would require approximately one-fourth of the working population to operate and maintain the phone system to handle the amount of phone calls we make today. The way the party line phone system worked is as follows: You picked up the black handset. All the telephones then were black handset phones that you leased from the telephone company. You could not buy your own telephone.

When you picked up the phone, the operator in Louisa came on and said "Number, please." You gave her the number you wanted, and she made a manual connection to who you wanted to call. If that line was in use, she would say that the line is busy, and you would have to hang up and try later.

When someone called you, your phone rang according to a code. Everyone on the party line could hear these rings. Some had a long ring and another could have a long ring followed by two short rings for example. Each customer soon learned his ring code. Our ring code was three short rings. Each customer had a different ring code. If the phone rang and it was not our code, then we knew it was not for us.

There were some people that didn't have much to do, so they would pick up the phone even though it was not their ring and just listen. This type of phone service was not very private, but there was nothing you could do to keep them from picking up the phone and listening in.

A long-distance call was quite expensive compared to today's standards. Toll-free "800" phone numbers did not exist.

Every time we had a thunderstorm, the telephone would get knocked out, sometimes for several days. Just like now, it was dangerous to be on the phone during a thunderstorm. I remember one time I was sitting at the end of our hall during a bad storm. The telephone sat on a small table in the hall about twelve to sixteen feet from where I was sitting. A bolt of lightning struck close to the house and I saw a red ball of fire about as big as an orange or grapefruit come out of the phone, drop down onto the floor, and start to roll towards me. This really scared me and I was thankful it went out before it got to me. Of course, the phone didn't work after this happened. One important usage the telephone was put to was to allow the civilian citizens to help defend the country.

About two weeks after the bombing of Pearl Harbor, December 7, 1941, my mother was approached by Mr. Brian Holladay and some others about setting up a aircraft spotter's post on the edge of Uncle John's field in front of our home. I was almost ten years old then. This spot was about 300 feet from our house. At this time, there was a real sense of fear in our country of being invaded by the Germans. People that lived in the outer banks of North Carolina and the Virginia Beach area could hear German submarines sitting out offshore, charging their batteries at night. The enemy came right up to the doorstep so to speak. As a result, the military came up with a plan to watch the skies and keep track of the airplanes that were in the air. Perhaps our area was picked for one of these tracking posts because it is now directly under a major Northeast/Southwest fly-way that jets use to go from Baltimore and the Philadelphia area to Charlotte and Atlanta. Several hundred planes a day go back and forth over this route. By having volunteer spotters in various places over the countryside, the military could keep track of the airplanes flying over our country.

All the local citizens were asked to take a shift to man the airline tracking posts so they could be manned twenty-four hours a day, seven days a week. There may have been some grumbling, but for the most part the people were willing to do their small part for the war effort. No one got any pay for the time they put in, nor did they get reimbursed for the gas to get there. The people, as a whole, were willing to make a sacrifice for their country. Never since has our country been as united as it was then. As Winston Churchill said, it turned out to be our "finest hour."

My mother had the job of scheduling the volunteers. To me, the worst shift was from 10 or 11 pm to 6 or 7 am, the night shift consequently, I never worked it. I remember working the post after coming home from school, relieving some wild-reputation girls and manning it up until supper time. Sometimes, I would man the post all day on Saturdays. If Mother couldn't find someone to take the slot, she did it herself. If someone had to miss a shift because something came up, mother would frequently fill in for them.

Mr. Brian Holladay was most instrumental in setting up this post, as he brought the military people to our home and talked my mother into taking the scheduling job. She had been a census taker for our district in 1940 and knew all the people in the area. Mr. Holladay also had a shanty placed on the edge of Uncle John's field and had a phone line run from our house to it. All the people (often two or more would man the post at a time) had to do was listen for an airplane.

The airplanes at that time all used piston motors and made a lot of noise, very similar to the noise today you hear when a helicopter flies over you. The airplanes then did not fly very high. I estimate they averaged two thousand feet high. There were no jet planes at that time. Now you only hear jet planes near the airports where they land and take off. When they get away from the airport, they get so high you do not hear them. Now airplanes can be tracked by radar.

You had a checklist sheet with numerous columns that had to be filled out for each time a plane went over. It stated, "Did you see or just hear the plane?" "How many airplanes did you see?" "How many motors did it have?" "How high up was it?" You had a picture selection in this column and you picked the one you thought best fit for the height. "How far away was the plane?" "What direction was it going?" It had more questions than this, but I cannot remember them all. After you completed the checklist sheet, you picked up the telephone and when the operator asked "Number, please," you said "Army Flash." If someone was talking on the phone, you simply butted in and spoke the phrase "Army Flash." When they heard that phrase, they were supposed to immediately hang up and then the operator would come in. After the operator connected you, there would be a little wait, and then someone would come on the line and tell you to go ahead. You first gave your post number, then you would read off the items on the checklist. You didn't hang up until the person taking the data told you to. Sometimes they would have questions. After you had phoned in your sighting, there was nothing to do until another airplane came along.

During these silent sky times, which might be two or three hours, you could read or do nothing, or tend the fire in the little stove in the shanty if it was cold. Some people listened to a small radio, but they had to keep the volume low so they could hear the airplanes. Usually the night shift was manned by two people, with one sleeping the first part of the night and later swapping roles. All the checklist sheets had to be kept, and every month or two a military person would come by and inspect the post and pick up the completed sheets. The people that ran the data receiving center must have kept up with all the airplane reports. One time I reported a family friend, Howard Shelton, who had a small single-engine Piper airplane, when he landed his plane in the long field right beside the spotter post. I reported the plane just like any other, and then I got fussed at and told I shouldn't have reported it. I didn't tell the center that the plane landed at our post. I think the plane stayed tied down in the field for several days. About one month later, we got a visit from the military asking questions about the single-engine plane that I reported. We told them what really happened and they accepted our answer.

Their interest and follow-up proved that the sacrifice the people were making was being used and not thrown away. They were keeping up with the airplane movements in the sky.

The airplane spotter post was just one of the things that affected peoples' lives during the war years from 1941 – 1945. Rationing of goods such as gas, rubber tires, and sugar were a few things that I remember were hard to acquire. Some things were impossible to buy, such as a new car. The U.S. Government controlled who got what iron and steel that was made and if you needed a new farm machine, you could order it, but you might have to wait a year or two before it was available. Everything was concentrated towards the production of things needed to win the war.

Today we do not realize how much war disrupts peoples' lives. Our leaders do not realize this, or else they would not be so quick to start a war. Even during the years of World War II, those living in the USA did not experience the horrors of the war that the people in

Europe and other countries had to live through. We did get a small taste of war on September 11, 2001, in New York City and how horrible that was. We should be doing things to change peoples' hearts towards war.

Chapter Four

Radio

For entertainment, we had the radio, magazines, books, and the Victrola. If you have never heard of a Victrola, it is a windup 78 RPM record player. We didn't have many records, but every now and then we would open up the Victrola. I liked to play records that had fiddle music. You played it by first putting the record on the turn table, then winding it up as far as the crank would go. It had a little lever that you pushed to start the turn table rotating, then you would very carefully put the pickup head with its needle over on the outside of the record. After the record finished playing, you had to remove the pickup head off the record, put it back in its resting pad, and turn the machine off. My older brother, W.O., had a friend, Marshall Price, who lived in Montclair, New Jersey. In 1939, Marshall invited W.O. to his home, and they went to the World's Fair which was in the New York City area. Marshall spent some of his summers at our house and he helped out on the farm. W.O. made a small personal record that I used to play sometimes on the Victrola.

I spent more time listening to the radio than any other form of entertainment. Before we got electricity, we had a battery, powered radio. Radios then were AM only. The stations we were able to receive were WRVA and WRNL in Richmond. At night, WWVA in Wheeling, West Virginia, came in really clear. You could use a lot of time going across the dial trying to find a good station. Today this is done with the television and a remote. Some things never change. Sometimes you would come across a station that came in really well, but in a little while it would fade away. We listened to WRVA most of the time. A small tall cedar pole about forty feet from the house was put up, and from its top a small antenna wire was run to the radio. A flat mesh wire strip was used where it ran under the window sill. If the antenna was not hooked up to the radio, you wouldn't receive any stations. Even with the antenna, sometimes the reception was really poor. Sometimes the radio played really clear for a while. This variation was due to the weather, as that affected how the radio played. This held true even after we received electricity.

The electrical-powered radio played much better than the battery-powered one, but still it was far from as good as the FM ones found today. Two of my favorite programs were "Lum and Abner" and "Amos and Andy." They had regular evening time slots, and you would try to schedule your chores so you could enjoy these comedy shows. People like to be made to laugh and that's what both of these programs did. They would make you laugh and you expected to get laughs from listening to them. I also tried to catch the news, and I followed World War II a lot on the radio. I very clearly remember on a Sunday afternoon about 4:00 p.m., hearing on the radio that the Japanese had bombed Pearl Harbor. Later a slogan was created which said "Remember Pearl Harbor."

The radio was in the dining room, and most of the time at breakfast and at supper, it would be on. We always liked to catch the weather report even though we were about as good at predicting the weather as the weather man was back then. When hay-cutting time came, it was sometimes a tough call as to whether rain was on the way or not. You needed a three- or four-day spell of sunny, dry weather to cut hay. After listening to the weather report, we would decide whether to cut hay or not. Back then they had no radar or satellites to get a picture of the storms that were approaching. Weather forecasting has greatly improved in my lifetime. My brothers liked to listen to baseball games on the radio, especially the World Series.

The first television I remember seeing was at the State Fair in Richmond, Virginia, in the late 1940's. The whole set took up a space about three to four feet in each direction. They were all black and white then. Even though the television had been invented, it was not until the 1950's before it was available to us.

We received farm magazines that I liked to look at. Another pastime was looking through the Sears & Roebuck, and Montgomery Ward catalogs. These two books had almost everything you could wish for in them. Some people called them the "wish books." I wished I was rich enough to send them an order that said, "Send me one of everything in the catalog." If such was ever done, you would have to have had an enormous warehouse to hold one of everything they had. For a lot of items, they would give you a choice of "good," "better," or "best." The "best" item was the highest price and the "good" alternative the cheapest. Most of the clothes and shoes we bought were ordered from these catalogs. We also ordered tools for the farm from these books. Now the internet has become the place to buy and sell, just like the catalogs. You buy from the convenience of your home just like you did out of the catalog.

We received the Richmond Times-Dispatch newspaper daily, brought to us by our rural mail carrier. This was for the weekday papers only, no Sunday paper. I read some of the war articles that were reported, but not everything. Some of the "funnies" (i.e. cartoons) I remember were "Dick Tracy," "Little Orphan Annie," "Terry and the Pirates," "The Phantom," "Blondie," and the "Kantzander Kids." This is just part of what was printed daily that I followed. My father read the editorial paper. The only thing I looked at on that page was the editorial cartoons. I spent more time reading the newspaper than any other single piece of literature. We also received the weekly *Central Virginian* paper. Then, as now, people would remark, "There's nothing in it." These were some of the ways we kept up with the outside world.

CHAPTER FIVE

BAREFOOT

Until I got to driving tractors, (thirteen –fourteen years old) every summer I would go barefoot, except when we went to church for which you always put on your best to show respect to God. When the weather got warm, usually mid-April or early May, you would begin to go barefoot. Most country kids did this. At first, it was difficult and painful because the bottoms of your feet were tender. You had to be careful where you walked outdoors. You soon learned to look for sandy spots in the road or grassy areas and tried to avoid hard and sharp places. After about one month of continuously being barefoot, the bottoms of our feet got tough enough so that rough surfaces did not hurt your feet. Even after your feet got tough, you always had to watch out for sharp things such as briers or anything else that could stick in your foot. You could get splinters stuck in your foot that would require someone to pick out or cuts that would require a band aid or wrapping of some kind. There are many hazards to going barefoot. One time I stepped on the tail of a moccasin snake where a straw rick had just been baled. I jumped back very quickly and didn't get bitten. When you are barefoot, you have no protection whatsoever against something like snakes.

Barefoot is dangerous and you always have to be aware of what you are about to walk into and consider maybe a detour around something you know is going to hurt your feet. If it is so dangerous and painful at first, then why do it? I say freedom or the feeling of no weights on your feet. It's comfortable, for your feet stay cooler except you better not step on a piece of steel that the sun has been shining on. Back then it saved money on shoes, so economics is another reason. If you came to a branch or creek, it didn't matter if your feet got wet, in fact, it felt good to walk in the water. Every night you had to wash your feet before going to bed and sometimes during the day also. If you walked in a cow pasture or around the barn working with the cows and were not careful and stepped in a pile of cow manure, you most certainly had to wash your feet before going in the house. The outside water spigot or water trough was a good place to do this. Going barefoot would usually end by October, or when it became cold and shoes felt better than cold feet.

CHAPTER SIX

Hog killing

Nearly every farm during the 1930's and 40's would raise several hogs for their own meat. It was a tradition that is almost totally gone now. How many people today know how to kill a hog? Hog-killing was done in the winter months, usually between December and February. Even during these months, you need to pick a cold time to kill hogs. When it's cold, you don't get as many flies. If it is too warm and flies are flying about, they might lay eggs on the meat you were going to cure. Several months later these eggs would hatch and larva (we called them "skippers") would form, spoiling the meat.

It takes several men working as a crew to kill hogs. Before the animal is killed, the scalding tub had to be set up and made ready. A scalding tub is a heavy open-top metal vessel big enough to contain a hog with room to spare. Usually a fire pit would be dug in the ground and the metal tub put over this pit. The pit needs to be longer than the tub, so that wood can be put under the tub at one end, and the smoke can come out the other end. The tub had to be three-fourths filled with water. A wood fire is built in the pit under the tub to heat the water until it was scalding hot. It was critical to get this water temperature just right before the hog was put in. This process would take two or three hours, so it was the first thing done on hog-killing day.

The hog would first be shot in the head with a small-caliber gun such as a 22-rifle. This would knock the hog out and he would fall down. A couple of people would then have to quickly jump into the hog pen and flip him on his back and, using a large knife, cut his throat. This would kill the hog and his blood would come gushing out all over the place. It was very important to get most of the blood out of the animal at this time. A board area was placed on one side of the tub with two log chains being laid across these boards. The dead hog was place on top of these chains. Remember, a dead hog probably weighs 200-300 pounds, sometimes more, so it takes several people to pick him up.

With the hog on the chains, a person would catch hold of one end of the chain, with another person on the other side of the hog to pick up that side of the chain. This set up required four people, as we would have two chains. The hog was lifted up and let down into the scalding water. The hog was left in the scalding water for several minutes, but not too long, and then was brought out with the chains and put back on the boards. Scalding the hog would make the hog's hair come off his skin easily. If the water was not hot enough, the hair wouldn't get loose. If it was too hot or the hog was left in the tub too long, the hair would "set" and then be very difficult to remove.

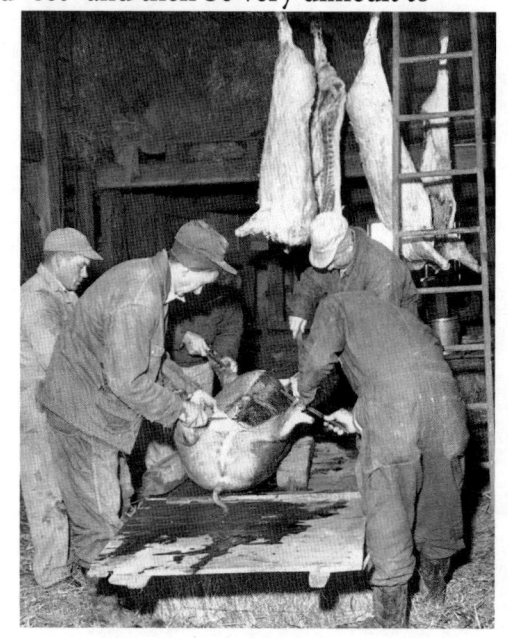

Everyone would pitch in and use their hands to pull all the hair off the hog. Some people would use round scrapers or other pieces of metal to get this hair off. What hair was not pulled off need to be shaved off with a knife. You need to get all the hair off the hog. The whole process was a very messy job. Talk about a dirty job, where is Mike Rowe from the program *Dirty Jobs* that we see on television?

**Pulling hair off hog after it has been scaulded to make the hair come off easier.
Courtesy of the Iowa Historical Society**

The next requirement was to have a long horizontal pole set up about eight feet above the ground. This pole needed to be strong enough to hold several hogs in a line. In the ankle area of the hog's rear feet are very strong tendons. A small slit is made on the back side of his legs and his tendons were pried open so a pointed stick called a gambrel could be put between its two back legs with the length of the stick being long enough to pry the legs open about eighteen inches. The hog was hoisted upside-down by two men (one on each side) with its two rear legs above the horizontal pole. A third person had to put the gambrel stick over the top of the pole and stick the ends of the stick between the tendons that had been pried out of the rear legs. Once this was done, the hog would be hanging upside-down with his snout a little way above the ground. The hog was then washed off and would be ready to be opened up.

Opened hogs hanging on pole so they will cool overnight. Courtesy of the Iowa Historical Society

A long cut would made in the hog's underside, a process we referred to as "opening the hog." A couple of wash tubs would be needed to be ready to catch the insides of the hog when this was done. When the hog was opened up, all the inside organs of the hog, including the heart, liver, kidneys, stomach, intestines, or guts, were put into these tubs. One of the odors I remember is coming home from school and they would be working on these intestines in the kitchen getting the fat off them. The smell was absolutely the worst. Some people cleaned the intestines out and then cut them up, cooked them, and ate them. They loved it. They were called "chitterlings." Usually, some black people would be helping with the hog-killing work and we would give the intestines to them, as well as the other parts we didn't want. I never could understand how they liked them. Of these organs, the only one we ate was the liver.

By this time, the day would be about over. The opened hogs were left hanging on the pole overnight, so that the meat would be cooled or chilled. It is important that the meat be cooled quickly after the animal has been opened else it might spoil. This was another reason for doing it in the winter. The next step was to cut out the meat. This was usually done the day after killing, but not always. One time, right after killing day, it got extremely cold, and the hogs froze on the pole. We had to bring them in the house and put them on sheets on the living room floor to thaw. We didn't heat this living room in the winter except on special occasions. On that occasion, it was several days before we got the meat cut out.

When you cut out a hog, you get hams, shoulders, bacon sides, pork chops or tenderloin, sausage, and fat, which can be boiled down into lard. When the cutting out was done, you have two categories of meat, fresh meat and cured meat. The fresh meat has to be eaten within a week or two; otherwise it would spoil unless canned or frozen. Until we got a freezer, the sausage and tenderloin spare ribs were canned. After we got a freezer, we froze the fresh meat. The meat to be cured, such as hams, sides, shoulders, jowls and heads was rubbed with table salt. You had to use plenty of salt and rub it into every crevice of the meat. In the latter years, we used Morton's sugar cure mixture which was still mostly salt, but not pure salt, as it contained spices. This salt and meat was put in a box large enough to hold the amount of meat you had. You put salt on the floor of the box, put in a rubbed piece of meat, and put more salt around it until you have packed all the meat parts with salt in the box. It was left in this salt pack for six weeks to two months, depending on the size of the meat.

When the meat had been in the salt long enough, it was taken out of the salt, cleaned off, washed, brushed off, and then hung up in the top of a small house called a "smokehouse." A short twine string was run through the edge of the skin in the shank area and then hung in the air on a stick or nail. It was important to not let the pieces of hanging meat touch each other. A smoke fire was built under the meat. You don't want a large fire that might cook the meat, as you are after the smoke effect. To get a proper "smoke," it takes a mixture of some dry wood and green wood, so the fire makes a lot of smoke, but little heat. You would do this every day for about two weeks. You never wanted to use pine. Pine wood smoke has an unpleasant odor. Each species of wood has its own unique smell or odor when burned and hickory wood smoke gives a better flavor to the meat. Other woods maybe used but they may impart a different taste to the meat.

After the smoking is finished, the pieces of meat were put in cloth bags. The bags were used to keep insects from getting on the meat hanging in the air. With the passage of time, the meat might mold, but this mold does not harm the meat. The longer you let this meat hang, the better flavor it would have. Two years is not too long for ham. I say it is an insult to the hog to eat the ham before it is two years old. We ate a lot of hog meat during these times, not realizing that all this hog fat will accumulate with time and cause clogged arteries and eventually heart attacks.

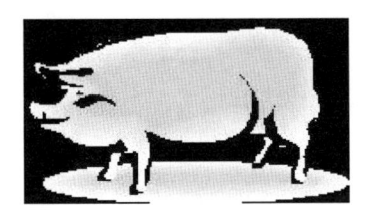

Chapter Seven

Food Refrigeration

In the time before electricity was available in the country, there were several ways to keep food products cool in the warm months of the year. One way was to use a spring box, a wooden box placed by or near a spring so that water flows into one end of the box and out the other end. This spring water was cool and flows 24/7. The drawbacks were that it is limited to how cold it would keep food, and it was usually not convenient, as the spring would be some distance from the house. However, it was much better than nothing. In our case we had one about three hundred yards from the house. This spring may have been used for water for the house before Daddy had a well dug at the end of the back porch. I was told that the first well he had dug came in dry. The one I remember was sufficient for our needs until the dairy barn was built and we had a lot of cows drinking from it.

Most every farm I know had an ice house. The ice house was made by digging a large hole in the ground which was about eight feet deep with steep sloping sides. The top of the hole was ten to twelve feet in diameter. A short building was built over the hole with an A shaped roof and one door. It had a ladder at the door that went to the bottom of the hole. A layer of sawdust was put in the hole followed by some straw. Usually during December to January, a cold spell would come that would freeze the water into ice of six inches or more thick, either on ponds or the river. This thick ice (the thicker the better) would be cut into blocks, loaded onto a wagon, and hauled to the ice house. The blocks of ice would be put down into the hole on top of the sawdust and straw. Additional sawdust and straw were put around the sides and on top of the ice blocks. Sometimes it would be August before these blocks of ice finished melting. The icehouse would be located near the house where it was convenient to put food to keep it cool. I think the icehouse kept things cooler than the spring box. One disadvantage to this icehouse was that it didn't last all year. With the warm winters we have now, there have been very few years that the ice freezes thick enough to allow ice to be cut into blocks to be used in an icehouse.

    I have a piece of furniture with ornate carvings on its front called an icebox. This icebox has a top that lifts up with the inside of the box being lined with metal. It has a drain hole in its bottom. The bottom is about six inches off the floor to allow a pan to be slid under it to catch any water that drains out of this hole. The water comes from the melting of the ice that is put in the upper compartment.

    In the town of Mineral, Virginia, was an ice plant that made blocks of ice about six inches thick by two feet wide by three feet long. It had a refrigerated pump powered by a big one-cylinder diesel engine like the one that powered the feed mill. The owners had an ice route that they ran in the summertime. They would load these big blocks of ice on a truck and go out into the countryside and sell the ice. They came by the house about once a week and would sell as much ice as your icebox would hold. You had to allow room for the containers you were going to put in the icebox, so you had to make an eyeball judgment as to how much to get. The delivery people had picks and other tools to cut the ice with. They handled the blocks of ice with ice tongs like logging grabs with sharp points to stick into the blocks of ice. We used the icebox much more than the icehouse, probably because the icehouse didn't have any ice in it year around.

    Before we got electricity, Mother got a refrigerator that ran on kerosene. This refrigerator was put in the house. You had to put kerosene in it and light it. I don't understand how a fire can make something cold, but I remember the frozen ice cream trays in the freezer compartment it contained. When we got electricity, my mother got rid of the kerosene refrigerator and got an electric one.

At one time a frozen food locker plant was operating in Louisa, Virginia, in the same building where the Louisa Auto Parts store is today. It had a large room with a thick, heavy door that was kept at zero degrees Fahrenheit or below. It was a large walk-in freezer with large drawers called lockers, as your key only worked for your locker(s). You rented these locker drawers that were about twice as big as a file cabinet drawer, and put meat or whatever you wanted to keep frozen in your locker. Daddy eventually bought a "deep freeze" freezer, which was put on the back porch, and we stopped renting the freezer drawer.

The back porch of the house got a lot of use. There we washed clothes, as the washing machine stayed on the back porch along with the wash tubs and the wash tubs bench. The cream separator and the milk buckets and strainer were all kept on this porch. The icebox was on it and when we got a freezer it took the icebox spot. Firewood was stored there during the winter months. Several chairs were kept on it as it was a good place to sit when the weather was mild. It was about eight feet wide by twenty-five to thirty feet long, but it served us well.

CHAPTER EIGHT

1940 Snow

Anyone that lived in central Virginia in January of 1940 will never forget the snow that started on January 24th at around three or four o'clock in the afternoon. By bedtime, it was several inches deep and still coming down hard. When we got up the next morning, it had stopped snowing. You couldn't see out the back porch, for the snow was plastered all over the screen. When the screen door to the outside steps for the back porch was finally pried open, one looked out on a world buried in snow. Some people said it was three feet. I think twenty to twenty-four inches was more accurate. It was difficult to tell because the wind blew so hard and the snow drifted so badly. It totally covered up most of the fences. Of course, the roads were totally blocked. We didn't go to school for two weeks. The milking, feeding and watering of the cattle and chickens took all day. We had some wood stacked on the back porch for our heat and cooking. Until the wind stopped, it was useless to dig out a path, as it would soon drift full behind you. After the first day, it got extremely cold and set records that still hold until this day. It went below zero degrees Fahrenheit every morning for five days, with one reading of minus twelve degrees Fahrenheit in Richmond, Virginia.

It took about fifteen minutes to get dressed to go outside after a significant snow. Back then a dress article was used called "leggings." I never see them anymore. Leggings were two separate heavy cloth or leather parts that were wrapped around the calf of each leg and held tightly by laces tied on the outside of each leg. They had a strap that ran under the shoe in front of the heel. This strap prevented them from riding up on your leg as you walked. The bottom of them was kind of bell-shaped and covered the top of your shoes. Their length was from just below the top of your shoe to just under your knee. They were shaped to fit the contour of your leg, being bigger in the calf area. To put them on, you first put an overshoe on your shoe. An overshoe is a flexible, rather thin rubber waterproof cover that you stretch over your shoes. They should fit your shoes snuggly; otherwise, sticky mud can cause them to pull off your shoe.

With your pants on and your shoes on with the overshoes pulled over them to keep your feet dry, you wrapped the leggings around your leg below the knee and hooked the strap under the overshoe. You gathered the pants leg cloth around your leg and made sure everything was inside the leggings, then put the laces over the eyes on the leggings and pulled everything tightly up on your leg. This tightness is needed so that the snow cannot get inside your shoe tops or your pants. You have to do each leg this way. It is why it took so much time to get dressed to go out in the snow.

Another way to keep the snow out of your shoes was to wear galoshes, rubber waterproof flexible shoe covers that are about twelve inches tall. They came up well above the top of your shoes. You pulled them over your shoes, just like the overshoes. Their tops were large enough for you to put the bottoms of your pants leg into them. You then pulled them snuggly around your leg and buckled them shut with their built-in metal buckles. Again all these various articles need to fit tightly to keep the snow out. You could put the galoshes on quicker than the leggings. Sometimes you had to get another person to pull them off when you had finished playing in the snow.

Yet another way to go out in the snow or mud was to put on gum boots. Gum boots are rubber boots that reach up to your knee. You don't wear any shoes with gum boots. You stick your foot with a sock or sometimes doubled-up socks into the boot and poke your pants leg down into the boot top. The gum boot does not have a seal around its top. A deep snow can go over the top of the boot and get down to your foot where it will melt and you will have wet feet. The main objection I had to gum boots was your feet would get so cold. Extra socks helped, but without the shoe, your feet could get very cold. They can be very hard to get off, requiring you to sit down and someone else pulling and wiggling them off your feet.

You also had to have a cap with ear muffs, coat, and gloves and if there was bright sun, sunglasses for complete protection so you could enjoy the cold outdoors.

When we did get our paths dug out, and I walked down the paths, the snow was as high as my armpits. After about one week, the snow got hard enough on top to support me. I would get up on top and proceed to walk on it; then all of a sudden, the crust would break and I would be up to my shoulders in snow. It was quite a struggle to get back to the paths. The river froze over solid and my brothers ice-skated wherever they pleased without fear of breaking through. If such a storm happened now, I wonder how much better we could handle it with four-wheel drive vehicles and the snow plows we have. I say it would still shut down everything, but maybe for not as long. I have heard that the weather will repeat itself in around one hundred years. Look out, year 2040.

Chapter Nine

Garden and Mill

Every farm had a vegetable garden. Some were bigger and better than others, but I don't remember a single family that didn't plant something. I think everybody planted potatoes. Depending on your taste, there would be English peas, snap (green) beans, butter (lima) beans, beets, corn, tomatoes, sweet potatoes, potatoes, spinach, turnips, lettuce, melon, and squash, to name most of the commonly grown vegetables. When the vegetables were ready to pick or harvest, a good portion of the crop would be canned in glass jars. This canning job was done on the kitchen stove. If you had a huge quantity, it could be taken to a cannery that had a larger and better canning facility that would can the vegetables into tin cans. Outsourcing the work to a cannery had another benefit of keeping the heat out of the house, as this work was done in the summertime. Most counties had a cannery located somewhere in the county sponsored by the County Extension Service that was run by the Department of Agriculture. After we got electricity and freezers became common, a lot of the garden foods were frozen instead of being canned. Freezing something is less work, and some things taste better from the freezer than from a can.

Along with canning or freezing the food we produced, another method we used to store food was to dry it. My mother used to dry apples. The apples were peeled and then sliced in pieces about one-eighth of an inch thick. These slices would be put on a clean wood board and placed on a set of trestle benches that were set up outside where the sun could shine on them all day. Each individual slice was placed on the board or boards until the whole board would be covered with thinly sliced apples. This was done in September or the early part of October. It would take about a week or more of direct sunlight to dry this fruit. The fruit had to be turned over so that the sun could get both sides. You could not let it rain on the half dried fruit. Every night it would have to be gathered up so that it did not pick up moisture or dew during the night. Thus, you had to collect and take the apples in every night. The pieces of apple would gradually shrivel up, and when totally dried would be much smaller than when they were cut off the apple. They would keep a long time when stored in a dry, cool place in the house.

You could eat the dried fruit directly out of the bag, but it would take some chewing. Soaking the dried fruit in water first would help, as it would soften them up. Mother used to make what she called "apple puffs," which was made by first soaking the dried fruit in water, then putting the pieces between two layers of pie dough and cooking it in the oven. The cooking caused the center to puff up. They were good, kind of like eating apple sauce in a pie.

Each summer we'd pick wild blackberries and raspberries. They were eaten directly or used to make ice cream. The ice cream you buy now is very mild compared to the homemade ice cream of my childhood which was made using real cream and fruit of some kind. Forget about all the fat it contained, this ice cream had some real taste to it! We had a strawberry patch and sometimes we picked wild strawberries. The wild ones had a strong taste. The huge strawberries you buy today may look good, but when you eat one, if you were blind, you wouldn't know what you were eating by taste. Not so with a wild strawberry.

There were two small orchards on the farm with apple and pear trees. The trees were never sprayed or pruned, so they didn't produce much fruit. We had two peach trees and a cherry tree. Most of the time, frost would kills the blooms on these trees and you wouldn't get anything that year. However, pretty much every year there would be apples. It took time to grow this food, gather it, prepare it, and preserve it. Remember that money was scarce then and when you used your time and labor to make food, it meant you didn't need as much money to live on.

We went to town, either Louisa or Orange, Virginia, about once a month to buy such things as cereal, flour, cornmeal, coffee, tea, sugar, yeast, baking powder, Jell-O, cake mixes, and other thing that were not grown on the farm. We ate a lot of corn bread made in an 8 x 11 inch pan. We called it "batter bread." Sometimes the corn bread would be made in a muffin pan. Biscuits were a morning staple. We also had rolls and some white loaf bread. Daddy called it "light bread." All these foods were cooked in a wood stove. Our wood stove was made by Home Comfort: but there were also other brands of wood cook stoves. Most of the cooking was what we now call "made from scratch," meaning you started with the very basic ingredients. Even though cooking was much more difficult and complicated compared to today, we never knew what being hungry felt like.

Another important component of country life in a community before automobiles and electricity was the mill. By the mill, I am referring to the place that had machines that would grind difference kinds of grains into a very fine powder used in the mixes that were put together and cooked that formed the very basics of one's diet. Bread is made from flour as the basic ingredient. The mills were usually powered by water. One hundred to two hundred or more years ago, the country, then just settled, was full of these mills that ground grains into meal or flour. When one drives around the country today, numerous places and roads have names that reflect the mills that were located on or near what the present names are for that particular place. The basic reason for so many mills was transportation. One needed to live within one-half day's walk or horseback ride distance to a mill. This would allow them to make a round trip to the mill in one day or less. They would take out of their storage some of the grain they had grown and take it to the mill to be ground into either meal or flour. They would take this corn or another grain (corn and wheat being most common) maybe once a month and get it ground, often paying for the service by trading some of the grain to the miller for his services. That way, they didn't need money, which they didn't have much of, so they exchanged what they did have, grain, in order to live off the land. We didn't use the local mill much, as we bought flour and meal at the stores in town.

The leftover parts from this grinding process, such as bran from the wheat, would be taken back home and fed to the livestock, pigs, chicken, and cows that enjoyed very much eating these parts. Having this done every few weeks allowed one to have a fairly fresh product to cook and eat.

One reason most mills were water-powered was the machinery then required a slow speed with lots of torque or turning force to rotate large stones that had grooves cut in them that crushed the grains into powder. By building a dam ten to twenty feet high across a stream of water and running some of this water over the top of a large diameter wheel, you get a slow-speed, high-torque power source that you didn't have to buy fuel for. Some metal was required for these mills, but they were mostly built out of wood or stone.

Holladay Mill Dam is in lower left side of picture. Courtesy of Francis Baker

Our community had one of these mills named Holladay Mills located on the North Anna River about three miles by road from our house. The mill is gone now, flooded by Lake Anna. It and the dam were located in Spotsylvania County about two hundred feet to the left when going north on Route 719 from Louisa County to Spotsylvania County. The pres-

Local musicians at Holladay Mill rehearsing for annual Galax Virginia Music Convention. Mill Dam in background with corner of mill in upper right corner of picture. 1920s NOTE: Hats and clothing worn in that era.

ent bridge across the lake is called Holladay Bridge. This mill was built by Mr. Brian Holladay, and he also ran it and served as the miller. When he stopped working the mill and the lake was formed, the mill was torn down.

Mr. Brian Holladay was quite a character. Sometimes he held Saturday night dances in his mill with live country music. He had many signs posted in and outside his mill. One I remember said, "Do not spit in the corner. Spit in the middle of the floor." I assume this was directed at customers that chewed tobacco, as they are constantly looking for a place to spit. Another sign I remember said "Pay no mind to the clown. He gets no pay for the entertainment." This is directed at himself as he could be comical. I would later use this phrase to sometimes describe the behavior of my son, Samuel Dick Harris, Jr.

GARDEN AND MILL

When one went to the mill, they did more than just get their grain ground. There would be an exchange of news and gossip to provide a break in the routine of the daily country life. When I was small, I heard my family talk about people who took all day to go to the mill when they could have made a round trip in two or three hours. They would stay and talk to whoever they met at the mill. I have never visited a modern electrical-powered flour mills. There is one in Culpeper about thirty-five miles north of Mineral. It makes flour from wheat shipped in by railcars. Nearly every day we see tractor trailer trucks from this mill going through Mineral, south to I-64 to customers in eastern Virginia.

The small community mills are all gone now, and we are dependent upon the large ones to grind our grain into food we can eat. It is difficult to realize that a good many people on this earth do not have enough to eat. We are blessed to live in a land with so much food. Now fewer and fewer people know how to grow a garden and preserve what it grows. If a great catastrophe were to happen (and don't say that it won't or can't happen), most people wouldn't survive. We don't like to think about such things, but there are no guarantees in life, except death and taxes. It's not pleasant to think about this, but it could come to a time when in order to survive, we would have to go back to some form of how life was before electricity.

It never hurts to have knowledge or skills of how to do things. You learn how to watch out for dangers, as life is full of danger. The problem with having only one specialized skill or narrow focus of knowledge is that life requires a variety of knowledge and skills. The Boy Scouts teach, "Be Prepared." When we depend on others to supply what we don't know how to do, then we become vulnerable when something happens that upsets the order of things. These catastrophes can be natural disasters such as volcanoes, earthquakes, etc., or political disasters such as war. We are powerless to stop them or to do anything but get out of their way. I hope you never have to live a survival existence.

CHAPTER TEN

SCHOOL

I began first grade in 1938 at Belmont in what is now the Belmont Community Center. The building is now owned jointly by the Belmont Ruritan club and the Belmont Womens club. The larger building that housed the high school grades in 1938, no longer exist. In 1940 Spotsylvania county consolidated all the high school grades in its new building at Spotsylvania Court House. The buildings were heated with wood stoves. They had no insulation and had large windows. When the wind blew you could feel some of it that came through the walls. With the wood stoves it would be hot near the stoves and cold near the windows. Most of the wood was stored in a separate shed near the buildings. Some was stored inside in the cloak rooms. There are axe marks in the floor to this very day, where this inside stored wood was split up. The only time hot weather was a problem was early September. The school year ran from the first week in September to the first week in June. If snow or ice closed schools, that time was made up by going on Saturdays.

When I began school, each person had his own desk. This was your desk the entire school session, unless the teacher moved you to another desk. This movement would usually be caused by two students in close proximity to each other that caused trouble. I never got moved, but I did see it happen. School hours were 9:00 a.m. to 3:00 p.m., with half hour lunch break at noon. We had a fifteen minute mid-morning recess and a fifteen minute mid-afternoon recess. This means we had five hours of study time five days a week. The recess time was used for going to the toilet and playing games or what is now called "Hanging Out" in small groups. We had two outdoor toilets, some distance apart from each other. One boys toilet at the far end of the school grounds, and one girls toilet near the school building. The early grade school subjects were reading, writing, spelling, addition, subtraction, multiplication and division. Hand held calculators did not exist then.

I went to Belmont for the first seven grades, then rode thirty miles one way to Spotsylvania High School for grades eight through eleven. I graduated in June 1949. I continued for one more year as a post graduate student as Spotsylvania, taking subjects I failed to take in the first four years there. For the first several years at Belmont you had to take your own lunch, and ate at your desk. Soon after the high school was moved to Spotsylvania, they had space to make a cafeteria, which they did and you had the option of buying your lunch at school or taking it from home in a lunch box.

One goes to school to learn the mentioned subjects, but the favorite time at school was recess. We all like to have fun. Group games we played were softball, boys baseball, a little bit of football for the boys, crack-the-whip, and marbles. We had a basketball hoop to shoot at, but no court. All these games were outdoors with the exception of marbles, which could be played either outdoors or inside. What is crack-the-whip? You get 10-15 people in a line to join hands. The biggest ones make up on end of the line. The whole line takes off running in one direction. After a short distance and the speed has gotten up, the large ones stop running on one end of the line and began pulling the line in a circle. Everyone holds on to each other very tightly. When one ends of the line stops and begins to pull the line in an arc, this causes the outer end of the line to pick up more speed. The speed gets so great that someone lets loose and some will go flying through the air and end up on the ground. It was fun for some, but you could get scuffed up or hurt.

Marbles was mostly a boys game. It is similar to pool in a lot of ways. A crude circle of anywhere form three feet to six fee or larger in diameter is drawn on a smooth flat surface either outside in the dirt or on the wood floor if inside. Each person puts in an equal number of marbles in the center of the circle. The pack of marbles is pushed tightly together like the pool balls are racked up. An order of shooting is determined. I forget how this order is determined, but each person in the game gets to shoot in a normal game. The object is to take a marble and knock the marbles in the center outside the line drawn on the surface being played on. A shooter takes a marble and holds it between the thumb and first finger, using either the right or left hand. Tension is held on the marble by the thumb and first finger. When the thumb is flipped the marble will be shot out at considerable speed and strike what it is aimed at. This first person to shoot has to break the pack. If by chance this breaking the pack happens to cause one marble in the pack to go outside the circle, he can continue to shoot as long as he knocks out a marble with each shot. When you miss a shot, it is the next persons turn to shoot. The game is over when all the marbles are knocked out of the circle. You kept all the marbles you knocked out of the circle, unless it was a free game. It was decided before the game was started if the games was for "Keeps" or for "Free."

To play this game you had to get on your knees. You would place your shooter marble between your thumb and first finger and put your knuckle against either the ground or the floor and up to the line drawn, but not over it. Some had the bad habit of when they were in the act of shooting, they would push their hand over the line. It was called "Hunching" and was a No-No. I have seen fights break out over doing this. The idle players kind of acted as referees for this game as they watched it being played. It took a lot of practice to become good at shooting. You had to keep your hand outside the line drawn. You had to aim the shooting marble very carefully at the one you were trying to knock out of the circle. Usually what happened, the one that shot first broke the pack and the marbles would be randomly scattered within the ring. The second person shooting would study the situation and knock the ones lying closes to the ring first. If you knocked out a marble with each shot, you continued to shoot until you missed or cleared out the ring. Like life you could lose your marbles very quickly if you were playing with one skilled in the game. This is the reason for the free game agreement before the game started, if he was likely to loose all his marbles. No one would play with someone if he knew would loose all or most of his marbles. In order to have a game if would have to be a free one. We did not play marbles at high school.

High school was considerable different from grade school. It was the consolidated high school for Spotsylvania county with a student enrollment of around four hundred. The building was less than ten years old when I started there. It was heated by central steam heat, but no air conditioner. The main difference was you moved from one room to another for each separate subject. Some subjects were required such as English, history, or math. Other subjects you had a choice of like typing, agriculture, shop, and home economics. Juniors and seniors had choices like chemistry, algebra, and geometry. I can't remember all the choices we had, but it was much more that what I have just named. We all had physical education or gym and had a football team, baseball, softball, and both girls and boys basketball teams. Many more sports than grade school. The school has a cafeteria and nearly everyone ate in it. Very few if any brought their lunch from home. The lunch break was at least forty-five minutes long, maybe an hour. You had five minutes to switch classes from one subject to the other. The halls were very crowded when the classes switched. Everyone remembers some things that happened when he went to school. As the years go by class reunions are set up and old friendships are renewed. We remember the good old days, yet when we lived those days we didn't think they were so good.

CHAPTER ELEVEN

DAIRY

From my earliest memories there were always cows to be milked, morning and night. Daddy and my two older brothers took care of these chores. Eventually, I learned to milk a cow by hand and had one or two cows to milk as well. They were very gentle and easy ones to milk, and as I remember they gave somewhere between one and two gallons each milking. The usual hand-milking equipment consisted of an open top metal bucket of approximately three-gallon size and a stool which was like a 2" X 4" piece of wood with a 1" X 4" to 6" wide flat top approximately 12" long nailed to the top of the 2 X 4. The stool would be approximately 12" to 16" tall. I don't remember any three legged stools that a lot of people associate with a milking stool.

Most of the time a cow was milked in a stanchion which was made of wood. A stanchion is what they call a device to hold a cow's neck that was made with two upright pieces of wood four to five feet long. One of these upright pieces was fixed top and bottom to four more short pieces of wood that came off each end 90 degrees to the long upright part. The other upright piece was pivoted at the bottom to go between the two 90-degree short pieces. When the stanchion was shut and locked, there would be approximately six to seven inches of space

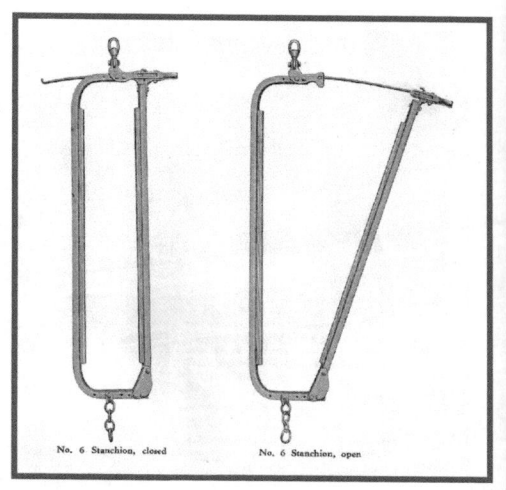

Cow stanchion showing both open and close position.

between the two parallel uprights. At the top of the stanchion, the two 90-degree pieces were twelve inches to sixteen inches long and when pivoted upright could open much wider at the top than the bottom. When the stanchion was open, a cow could lift her head a little and step forward to put her neck between the two uprights. A person then had to walk beside or in front of the animal and push the stanchion shut. When the stanchion was wide open, one end of the short block of wood on a pivot was set up on top of the pivoting upright. This pivoting upright was longer by about two inches than the fixed upright. When the upright was pushed to the side of the cow's neck in place, the short pivoting block would drop behind the pivoting upright and lock it in place so that the cow could not pull her head back and get free. This describes a homemade stanchion.

To do what we called "turn a cow out," a person had to lift up this short pivoting block and push the pivoting upright to the side, which opened the top up enough so that a cow could back up and be free of the stanchion. Most of the time, the above process was very routine, but it would become quite a hassle to catch a cow in the stanchion if she didn't want to be caught. Usually some feed, either hay or ground grain (called "chop") was put in the manger. The manger is what is called the area below and in front of the cow where she can eat. The Baby Jesus was placed in such a place after he was born.

When a cow didn't want to put her head in the stanchion, first you made sure some smelly feed was in the manger. Then a small rope or heavy string was tied onto the pivot upright so that a person could stand eight feet to ten feet to the side and pull the stanchion shut once she put her head in. These stanchions were usually in line with four to five feet between them. It helped a lot to get several gentle cows caught and eating in their stanchions and then get the wild one to calm down and see what her sisters were doing. Then you could gently urge her to step into the empty stall and begin eating like the rest. When this was done, then a person pulled the rope from the other side of gentle cow and locked up the wild one. You better hope the stanchion was made strong enough to hold her, as there would be much jerking and pushing trying to get free. Sometimes it would take a week or more to get a new animal gentle enough to go in a stanchion like the rest of the herd.

My father originally built a barn with a hay loft and horse stalls on each side with a wide manger down the middle and one harness room. When he first started milking cows, he built a shed onto the horse barn, and then built a second shed onto the back of the barn. Then he added additional sheds onto these. In other words, he kept adding to what he had over the years; as he acquired more animals he would add on sheds as needed. He put stanchions in these sheds until he had about twenty. Of course, all the sheds had dirt floors. These floors would be bedded with uneaten hay or straw from the straw rick left from the threshing machine. We called this complex of sheds and horse barn "the stable." Sometimes the very gentle cows were milked out in what we called the "stable lot," which was outdoors. This stable lot was a small fenced area that the animals could roam around in. The lot contained the corn house with grain bins and a shed, the buggy house with a wagon shed, and two pig pens. A watering trough was filled from a well off the back porch of the house. The water was pulled out of the well by two buckets on a chain by hand. When the nearly, filled bucket was pulled up, you were sending the empty one down to be filled. The buckets were dumped into a tub that had a pipe about fifty feet long that ran downhill to the cement watering trough at the edge of the stable lot. The cows and horses drank from this trough.

Before a cow is milked, the bottom sides of her udder and teats should be washed off with warm water to remove any trash or manure which could fall into the open bucket. As I remember, most of the time we used a dry rag to do this unless the cow was extremely dirty. After the grade "A" barn was built, the warm water was used 100 percent of the time.

It didn't matter if the cows were clean or dirty they all got washed before being milked. When this procedure was done, it causes the cow to do what we called "let down her milk." If the cow doesn't let down her milk, it is very difficult to get the milk. In fact, you won't get very much.

Washing the cow's teats prior to milking.

A cow is milked by hand by sitting on a stool with the bucket between your knees and catching hold to her teats and squeezing and tugging at the same time on the teats, if you do it right, the milk will squirt out with force into the bucket. You had to put your head into the recessed area called the flank between her leg and large stomach. By doing this, you can feel any coming movement the cow is going to make. A cow can kick very quickly and if you don't move fast, the cow will kick the bucket or step into it. If the latter happens, you dump out any milk that is left in the bucket, wash it out, and start over. If they just kick the side of the bucket, usually the worst that happens is that just some of the milk is spilled out. If a cow gets sore teats, usually in the winter, then you better be very careful and watch out for them kicking. This why Bag Balm ointment was invented and to this day serves as a very good salve for both animals and humans that have flesh cracks that need to heal. Usually several cats would hang around hoping to get some fresh milk. Sometimes you would be tempted to tilt the teat sideways and squeeze to gently squirt the cat from six to ten feet away. They didn't mind; they would just lick themselves off.

At this time the milk was separated and the cream was sold. The skim milk that was left over was mixed with ground grain. This concoction is what we called "slop," and we poured it into the trough for the pigs and hogs. When the pigs realized that they were about to be fed this slop, they would jostle, push, shove, and squeal. It was pure bedlam at the trough to get to this milk slop. Talk about manners! This is why sometimes people at a food table are compared to pigs, in that a person does not restrain himself, but acts like the pigs at the feeding trough.

After the cows were all milked, the milk was carried to the house in the buckets where it was poured into a machine which separated the cream out and left the skim milk. We called this machine the "separator," and it was kept on the back porch. The machine had a bowl on top that would hold maybe three to four gallons. The bowl had an adjustable spigot on its bottom. When this spigot was opened, the milk flowed into a stack of rapidly spinning cone-shaped discs. The machine had a handle that a person cranked with his arm that caused these discs to turn. The discs spun so fast that they caused the heavier cream to separate from the lighter part of the milk through centrifugal force. There were two outlets or spouts from the disc cage, one for cream and the other for skim milk, which is like one percent milk found in the stores today. You had to have two containers or buckets to catch what came out of the separators: one for the cream and one for the skim milk. The cream was put into a metal can, about six-gallons in size, with a top. This can was set in a tub of well water to cool and to keep it cool. As mentioned, the skim milk was given to the pigs, hogs, and sometimes the chickens.

A pickup truck from Orange came twice a week and picked up these cans of cream. They then transported them to the Orange Creamery. What they made the cream into, I do not know. I would think it was used to make butter, as ice cream was not common then. Of course, we had plenty of whole milk to drink morning, noon, and night. Mother made butter in a hand-churn about every two weeks. All this high-fat dairy products is what we now know causes clogged arteries and heart attacks.

After the milk was separated, all the parts of the separator had to be pulled apart, washed, and rinsed, including the milk buckets and strainers. This entire process of getting the cows in the lot, catching them, feeding them, milking them, turning them out, separating the milk, and washing everything up had to be done both morning and night every day, including Sundays. This really runs into a lot of time and is sort of like prison, in that once started, you cannot just shut it down and start up again at a whim. When a person got sick, somebody had to do the work that person normally does. Such is a dairy person's life.

Daddy started selling what was called grade "B" whole milk to the Farmers Creamery in Fredericksburg in March of 1941. A nearby neighbor, Billy Goodwin, got a pickup truck and started a daily milk truck route to Fredericksburg. He hauled our grade "B" milk along with several other neighbors' as well. Grade "B" milk was whole milk that was not produced in inspected, clean, state-of-the-art facilities. This classification also included milk that was not quick-cooled, or properly cooled. The night milking was put into ten-gallon metal cans and set in a tub of well water. This cooled it down somewhat in the summer and worked very well into the winter. No attempt was made to cool the morning milk, for soon after the morning milking was finished, the milk truck picked up both the night before and the morning milk, still warm from the cow. The milk arrived in Fredericksburg later that morning. When milk comes from a cow, it has the temperature of the animal. Cows and humans have about the same body temperature. If you drank this warm milk soon after it comes from the cow, it tastes totally different from normal processed milk. If milk is not cooled or processed soon after it comes from the cow, it will spoil and become unfit to drink very soon. In the summer, the milk can go bad within twenty-four hours. If it is cooled soon after the cow is milked to below forty degrees Fahrenheit and kept below this temperature, it will keep at least one week or longer. It is very important to keep milk cold, but not frozen.

I think it was when we started selling this grade "B" milk that Daddy decided to build, what was at that time, a state-of-the-art milking facility. He got blueprints from Virginia Polytechnical Institute, now known as Virginig Tech, for a thirty-cow dairy barn with a three-room milk house. If he built this facility, purchased milk-cooling machines and everything passed inspection, he could sell the milk as grade "A" milk. Grade "A" milk brought a better price than either grade "B" milk or cream.

Dairy barn and milk house. 1943

In the early spring of 1943 (either February or March), construction began on the new barn and milk house. Mr. Charlie Carpenter was the main person on this project, as he was skilled in construction of blocks and cement work as well as with general carpentry. Daddy hired helpers to assist Mr. Carpenter and took on the job of getting the materials in place to put up these buildings. Daddy thus served as his own contractor. The barn and milk house were built out of cement blocks acquired from Mr. Duggins in Louisa County. Daddy acquired the sand from sand bars in the nearby North Anna River. He acquired the rocks from various rock piles for a base for the cement. Last, the cement bags came from C.R. Butler in Orange. All the ditches for the foundation were dug by hand. A layer of rock would be put in the ditch, and then the sand and cement mixture was mixed by hand in a shallow box and then poured over the rocks. The walls were made by the laying of cement blocks. All this was heavy work, done by hand. The floors were all cement, with the barn floor having several contours to it. Next to the outer walls was a feed alley, then a curved manger, then the stanchion curb with steel pipes to hold the stanchions. Next were the cow stalls area, and last a gutter to catch the cow manure. This describes one-half of the floor plan; the other side had the same plan. The center of the barn was an eight-foot-wide flat floor between the gutters. There were fifteen stanchions on each side.

The outer dimensions of this barn were thirty-six feet wide by sixty-two feet long. The barn had an upper story, the loose hay loft. Some people called it a "hay mow," but we always called it the loft. It had a hay-fork track system in the very peak of the roof. The ceiling to the milking area was formed by the floor of the hay loft and was made of two-inch-thick oak joists with a pine tongue-and-groove floor for the hay loft. A smooth asbestos ceiling board was nailed to the underside of the hay floor joists. The rafters and sheeting were pine that was cut on the farm, as well as the oak. Both of these were taken to a sawmill and cut into timbers in the size the plans called for.

By about mid-June 1943, the barn construction had advanced to where the roof was completed and the loft floor was in and ready for loose hay. I don't remember who came up with the idea to hold a Saturday night barn dance in the newly-built hay loft. It was decided to hold one before any hay was put into the new loft. Within about one week after finishing the hay loft floor, a dance was held with the community being invited. A temporary rickety ladder was made for people to get up into the loft, about ten feet above ground level. I remember some of the people were afraid or hesitated to go up the incline as it swayed and moved when people went up it. They were afraid it was going to fall down, but it didn't. I think we had somewhere around 100 people come to the dance. Remember, this was in the midst of World War II (1943), and gas and rubber tires were rationed and lighting was furnished by oil lamps and lanterns. They worked well enough to give enough light for the dance. Unlike today, in construction, electricity was not used to power hand tools. No electrical tools were used in the barns construction. Everything was done with hand tools, and since the building was only partially complete at this time, it had no electricity. The electrical wiring was one of the last things done in its construction.

There was live music by Lewis Talley, who played a fiddle, accompanied by Haywood Talley and Franklin Kean, who both played string instruments. The musicians who played at these types of dances were paid by what we call in church "love offerings." A hat or container was passed around about midway through the dance and people would give whatever amount of money they wanted. This was then given to the music makers for payment for their services.

While at the dance, I remember listening to Albert Bazzanella explain to someone how the rafters were made. Albert lived across the river from Cherry Grove and was a very skilled carpenter. He had worked on the Pentagon just outside of Washington, DC. Daddy got him to come over and work with Mr. Carpenter who at the time was having trouble reading the blueprints. Albert understood what needed to be done and worked with Mr. Carpenter on making a "jig" or pattern to construct each half of the rafters. After, all the rafters were made and put up, Mr. Bazzanella went to another job. I also remember roaming around outside among some of the cars and listening to some of the talking and laughter among those that were drinking some type of spirits, right out of the bottle. Dancing and drinking seemed to just go together. For a very short time, we had a perfect place for entertainment for the people of the community, and we sponsored a dance party for them. This was what you would call a genuine "barn dance." Everyone had some fun. The very next week we started putting loose hay in this loft, using the hay-fork system that was put up in the top of the loft.

The milk house was a small twelve-feet by thirty-feet one-story building with three rooms. It had a small boiler room with a chimney in which a small wood-fired upright steam boiler was set up. It was hooked to a water line that ran up to the house and hooked into the house water system. This boiler provided hot water and steam for washing the milking equipment. The next room, the wash room, was twelve-feet by twelve-feet and contained a two-compartment wash tank and a steam cabinet. This steam cabinet had two levels made by three-quarter inch pipes with small holes drilled in the top of the pipes. The ten-gallon milk cans were placed upside down on the bottom rack of pipes after they had been hand-washed and rinsed. The upper rack was used for milk pails, buckets, and strainers after they had been cleaned. When the steam cabinet was loaded with milk equipment, live steam from the boiler was turned into the cabinet. The steam came out of the small holes and shot up into the inside of the cans and buckets that were turned upside down and rested on these pipes. The object was to sterilize or kill any germs on the milk equipment. I always questioned if the 180 degrees Fahrenheit temperature (that the inside of the cabinet reached) actually killed anything. The secret to having clean milk equipment was to rinse off any milk on the metal as soon as you finished using it. If you let the milk dry on the metal, then it would take a considerable amount of hand scrubbing to get it clean.

The milk room was twelve-feet by twelve-feet and contained a box cooler made by International Harvester and was powered by an electric motor that ran a compressor. This cooler could hold about six or eight ten-gallon milk cans. On the inside of the box cooler were a series of pipes. These pipes ran around the inner walls with several inches of space between the walls and the pipes. Several hundred gallons of water were put into the box, which had a large top that opened up. The compressor pumped the refrigerate through the pipes which caused a bank of ice about six inches thick to freeze around all the sides of the box with the pipes being in the middle of the bank of ice. The bank of ice kept the water in the box very cold at all times. For quick cooling of the milk after it came from the cow, the milk ran over a device called an aerator, made by a series of pipes set on top of one another. These pipes were joined together to make a smooth ripple surface (both front and back) over which the milk flowed in a thin sheet. The very cold water from the ice bank cooler was pumped through this series of pipes by the electric-powered pump, beginning at the bottom and going back and forth inside the pipes. The milk then flowed over the outside surface of the pipes, which in turn, caused the milk to rapidly cool. As a result, the formerly cool water became warm water and it flowed back into the ice bank box through the top pipe. However, just like a home refrigerator, the compressor would run between the milking cycles which would then, in turn, cause the water to cool back down between milking.

In my opinion, the freshly-cooled milk that came off the aerator through this process had an absolutely delicious taste, in comparison to the milk produced by present milk handling standards. If dairy owners and operators want to get back or increase the milk-drinking market today, then they need to figure out how to duplicate this taste in their product. People desire and will pay for outstanding taste no matter what the food product is. Whether it is steak, bread, fruits, or milk, people want outstanding taste and quality.

The aerator was basically a rack of pipes that hung from the ceiling of the milk room. In terms of dimensions, it was about three-feet wide and eighteen-to-twenty inches in height. The aerator had a tank on top that functioned like a cream separator bowl, (except that it was much larger) into which the milk was poured. It had an adjustable spigot in the bottom of the tank. This spigot was partially opened and routed the milk into a U-shaped trough which consisted of tiny holes spaced about one-inch apart. These holes ran the entire length of the aerator. The milk came out these small holes and contacted the cold pipe, resulting in the milk spreading out into a thin film from front to back. This film of milk clung to the ripple surface of the pipe until it reached the bottom of the pipe where it then dropped off into another U-shaped trough. The bottom trough contained one center hole and a swinging spout that directed the milk into the ten-gallon milk cans.

The ten-gallon milk cans sat on the cement floor. When the can was full, the swinging spout was quickly switched by the operator to the next empty can. One of the problems with this process was that if the operator wasn't paying attention, the milk cans would overflow, thus wasting the milk. When this happened the operator, or someone else, would frequently let out an oath because of the wasted milk. Next, he had to use a cup to dip out the milk in the overfilled can to the proper can fill level. After the milking was finished, all of the metal

Ten gallon milk can. Courtesy of the Iowa Historical Society

parts were washed, scrubbed, and rinsed to keep them clean. The night milk shift cans had to be set in the ice bank milk cooler to keep them cool. The morning milk cans were loaded directly onto the milk truck when it arrived which was normally soon after the milking was finished.

A gallon of milk weighs roughly 8.6 pounds. A full ten-gallon milk can would weigh over 100 pounds, 86 pounds being the weight of the milk itself. The cans had two handles, one on each side of the tapered part of the can. The can's top had a slightly-tapered ring about three inches wide that would fit into the straight neck of the can. The can's top was typically banged into place with a wooden block so as to prevent it from coming off while one was handling the full can. When the milk truck hauled several different farmers' milk cans, the driver had to keep track of whose cans belonged to whom. This was accomplished by each customer having his own unique can number. Our can number was 57, although I do not recall any of our neighbor's numbers. In addition, once the milk arrived at the milk plant in Fredericksburg, the workers at the facility had to maintain a record of whose milk it was and how much milk came from each dairy. The can-numbering system enabled this.

As a dairy farmer, one of the things we had to look out for, particularly in early spring, were wild onions. The cows were let out to pasture to graze on grass, and they would eat wild onions. This would cause the milk to smell like onions which, in turn, would cause it to be rejected at the milk plant in Fredericksburg. The milk would be returned and would have to be discarded. Another thing I learned was to get the cows out of the pasture by noon and feed them hay. This would allow the onions to abate out of their system before they were milked.

Milk was paid for by the hundredweight, as it still is today. Of course, the rate has changed over the years, but not nearly as much as the expenses incurred by the farmer. In the 1940's, a farm tractor cost anywhere from $2,000 to $4,000 dollars. Today, a tractor costs twenty times that amount, but the milk price has only increased two or three times over the past sixty years. The present day dairy farmer exists in a very tight economic situation. Survival, much less making a profit, is extremely difficult. The dairy farmer's situation has changed dramatically from what it was in the 1940's and 50's.

As previously mentioned, one of the important steps in milking is to clean or wash off a cow's teats and udder before beginning to milk. Another important step is to check each teat's milk in a strip cup for clots in the milk. A strip cup is a small metal cup about the size of a coffee cup. It has a funnel-shaped top with a flat mesh screen on the bottom of the funnel. You milk several "tugs" onto this screen, and if you do not notice any clots or other abnormalities, then the milk is fine. Each teat is done this way. If the teat's milk does not pass, then the cow has to be milked and that cows milk has to be kept separate from the good milk. It is either fed to some animal that it will not harm or thrown away.

DAIRY

After the cow has been milked, the milk is poured into a strainer. The strainer contains a flat bottom and is full of small round holes. It is made to fit into the top of a milk can. A gauzed-back thin cotton filter is placed onto the bottom of the strainer. A round metal piece containing small holes is then placed on top of this cotton filter and held in place by a round spring ring. This ring holds everything in position so that when the milk is poured into the strainer, all of the milk goes through the filter system. The filters are then disposed of after each milking session or if they become clogged, and replaced with a new one. The object of the strainer is to catch any trash or debris that may fall into an open bucket of milk. Unlike the cream separator with its strainer filtration system set on the top bowl, the grade "B" milk system has a strainer on each milk-can. The aerator also contained a strainer set inside the top tank. The key is that the milk was always ran through a strainer, be it cream or any grade. Modern milk pipelines contain a section for similar filtration systems.

When the milking was performed in the stable, the cows were milked by hand. The cows were still hand-milked when we moved from the stable to the new barn in either September or October of 1943. As I recall, it was around December of 1943 when we purchased two International Harvester milking-machines. A milking-machine works by the use of a motor-driven vacuum pump to create suction. In addition, the milking machine consisted of a pail to catch the milk, several rubber or plastic hoses, a pulsator, and four teat cups. A teat cup consisted of a metal outer shell which contained a rubber liner. The rubber liner was

Teat cups hooked to a pail-type milking machine ready to be put onto the cow in a conventional dairy barn with stanchions.

Milking maching pail-type - milking a cow in a conventional dairy barn. Courtesy of the Iowa Historical Society

connected to a claw with all four cups being attached to one claw. One milk line ran from the claw to the milk pail or to a milk pipeline. Another small vacuum line runs from the pulsator and connected to the metal outer shell. When suction was pulled on this line, it caused the rubber liner to simultaneously squeeze and tug the cow's teats just like a person would by hand. The pulsator turns the suction action on and off in a two-by-two motion. A person would use two hands while milking, one of which is used to squeeze and tug while the other hand relaxes and gets in position for the next operation. Thus, two teats are alternately milked. However, the pulsator performs a two-by-two hand motion using two small cups at the same time (as opposed to alternat-

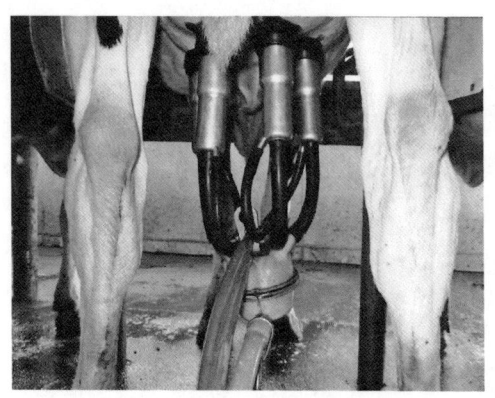

Close up of milkers hooked to a cow. Milk can go into a pail or into a pipeline.

ing). While two teats are squeezed and tugged by the vacuum, the other two are relaxed waiting their turn. They go back and forth during the milking process as the pulsator makes a "chug-chug-chug-chug" noise. If you placed your finger into a empty teat cup while in operation, you could actually feel the squeezing and tugging action.

Being able to tell when the machine was finished milking the cow was a learned operation. One way to tell was to pinch the hose to feel if there was any milk flowing through it. If the hose was clear, you could actually see the milk flowing. Last, you could feel the udder and tell when the milking process was finished. Today, milking machines have automatic shut-off systems. These modern systems work through a detection device that determines that milk is no longer flowing, which in turn, triggers a line to move and remove the set of milk cups as well as shut off the vacuum.

Receiving electricity in the latter part of 1941 (just prior to Pearl Harbor) made the construction of our then modern dairy facility in 1943 a reality. With electricity, we could pump water into the dairy, perform the necessary cooling operations, run the milk machine vacuum pump and, of course, run electric lights. Until then, we were forced to use kerosene lanterns which produced poor light. These lanterns were hung on nails in the shed's loft timbers. Common sense dictates that lanterns are not the safest thing to have in a barn around hay and straw, but prior to electricity, we had no alternative. The advent of electricity provided for advancements in the quality of farm life that we could only dream of and allowed us to have a modern dairy. This modern dairy allowed us to expand the farm in other areas and acquire state-of-the art machines. Dairy and farm life was still hard work, but our milk production greatly expanded once we acquired electricity. Other than electricity, two additional factors had a tremendous impact on the quantity of milk that was produced: advancements in genetics, and artificial insemination.

Milkers in a modern dairy set-up or parlor. Milk goes from the cow into a pipline, then is stored in a large bulk tank that keeps it cold until it is picked up by a large tank truck.

In 1947, we joined the Dairy Herd Improvement Association, or DHIA. Under the DHIA, each cow's milk was collected and weighed both morning and night. This was done one day of each month, and also a small sample of milk was collected and tested for butterfat content from each cow. A DHIA representative ("the milk tester") would come and collect the samples and the weight data. DHIA members paid a fee for this service. When we joined the DHIA in 1947, the herd average was around 8,000 pounds per animal per year. This means that the average cow in the herd would produce 8,000 pounds of milk in one year. The normal length of time a cow is milked in one year is ten months. This provided the cow with a two-month rest period, during which time they are referred to as "dry cows." Since this was an average figure, some cows produced more and some less. By tracking the production record of each cow, the farmer has proof as to what each animal is producing.

In 1947, it was rare for one cow to produce ten-gallons of milk in a day, as we normally had only one or two animals that did this at a time. Today, the same herd produces 24,000 pounds per animal as an average with most animals producing ten or more gallons a day as the norm. One of the primary reasons for this tremendous increase in herd average is advancements in genetics and the advent of artificial insemination (or "AI" for short). AI allows for the genes from the above-average ten-gallon per day cow to be passed on to thousands. This is accomplished by separating the bulls from the high-producing cows, and collecting and freezing their semen. These samples are then diluted and shipped all over the country so that one bull can be used to breed thousands of cows. This was not possible prior to the advent of AI.

Through AI, the dairy farmer could triple the production of an animal. What would naturally take several generations of selective breeding could be done very quickly and easily to get these dramatic increases in production. However, with this great increase in production, the law of supply and demand comes into effect, as an excess surplus of milk means lower prices to the farmer. In the 1940's and 1950's, the areas around major metropolitan areas were full of thirty-cow or more sized dairies. Today, there are very few farmers that milk only thirty cows. "Factory" dairies, now milk over 1,000 cow herds and are replacing the family dairy farm, which typically milk 100 cows. The small dairy farms that were typically at least thirty or forty miles from big cities have been replaced by suburban sprawl. This tremendous change in milk production in our country over the past sixty years is just one of the ways life has changed.

The Virginia Department of Agriculture and Consumer Services keeps records of the farms in the state of Virginia and what they produce. For dairy farms, the records show there were 2,700 dairy farms in Virginia in 1976 that produced 2.1 billion pounds of milk. In 2012, the number of dairy farms in Virginia had dwindled to 647 but had produced 1.7 billion pounds of milk. These records tell us that the amount of milk produced in the state of Virginia had decreased from 1976 but not nearly as drastically as the number of farms that produced the milk. This is proof of the changes occurring on dairy farms.

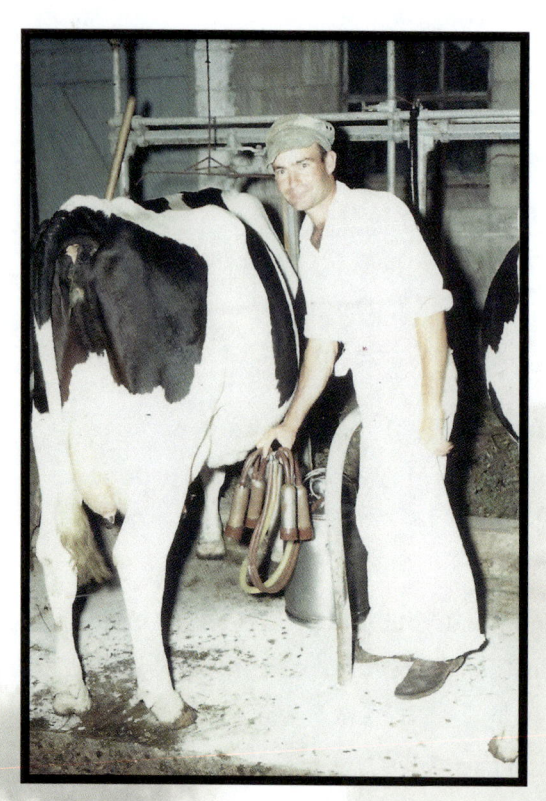

LEFT Brother Fred preparing to put the milkers on a cow.

BELOW Brother W. O. washing a cow's udder before milking.

Grain binder being set up to cut wheat.

Bulldozier building a street in the Town of Louisa operated by H. T. Barton.

Raking alfalfa hay with John Deere "B" driven by Anderson Williams.

Pedro Bastos loading a bale of hay on a wagon. Jimmy Downing on tractor and author on wagon.

CHAPTER TWELVE

CORN

The field crop that took the most work, by far, was corn. First, the field that was to be planted in corn was plowed with a turn plow. Before we purchased a tractor in late November or December 1940, this plowing was done with a team of two horses and what was called a "walking plow." The walking plow had one single moldboard, that curved part of the plow that causes the dirt to be turned over. It would take a person with a plow and a team of horses or mules all day to plow one acre if he was lucky.

Farm tractor pulling a 3-bottom trailer plow. Courtesy of the Iowa Historical Society

This plowing was done whenever the weather would let you during the winter months. The earlier you could get this done, the better, for the freezing and thawing of the plowed dirt would cause it to crumble up easily in the spring. One thing you did not want to do after the freezing was over was to plow the dirt when it was wet. When you work dirt when it's wet, it will dry out into very hard lumps. This is especially true of the red dirt we had which had a lot of clay in it. This red dirt is what is used to make bricks. If you work it when it is wet, you will be trying to plant in what amounts to broken-up bricks. Farmers referred to them as "clods."

After the field was plowed, the next operation was to use a heavy drag or drag harrow about eight-feet wide, over the fields. This knocked down the high spots left from the plowing and made the fields considerably smoother. A drag harrow, a more refined version of a heavy drag, was built in a V-shape out of heavy timbers. It had holes drilled through these timbers and long steel spikes like a railroad spike were driven into these holes. The spikes stuck through the timbers about four inches and would rip the dirt more than the smooth drag.

After dragging the field, the next operation would consist of laying down lime over the whole field. The lime would typically be ground limestone, a very heavy material, ground into a dust and handled in bulk. A dump truck would bring it to the field where it was dumped in a pile. A lime spreader was used to spread the lime over the field. We would receive a freight car load (fifty tons) of lime at a time which was delivered to the nearby train depot in the town of Mineral. This freight car was a boxcar, which is an enclosed train car with doors on each side in the center of the boxcar. The railroad would allow you two or three days to unload the fifty tons of lime out of the box car; otherwise, they would charge you extra for the freight cost. If it took you more than a week to unload the boxcar, it would cost you more for the freight than for the actual lime.

The boxcar would be parked off to the side below the depot beside what is now the Mineral Fire Department's lot. We would pull our truck up real close to the open doors and used a long-handled shovel to load the lime onto the truck. Lime is very heavy and dusty. After the truck was loaded, the top of the load of lime had to be watered down with a water hose at the depot. This would keep the lime from blowing off the truck as you drove down the road. You needed to get the lime hauled and spread on the field between rains. On a farm, the weather is always a big factor in whatever you do. You learn to work with the weather.

A lime-spreader was a two-wheeled (one on each end of the machine) V-shaped box with a series of adjustable holes about eight inches apart down the entire length of the box. An agitator bar was mounted over the top of these holes, it was turned by the wheels on each end of the box. The spreader was backed up to the lime pile and loaded by hand with shovels. The spreader was then driven away from the pile a short distance and the adjustable gates were opened. Adjustable gates could be used to set the amount of material you wished to spread onto the field. When the gates were opened by a lever, you continued to drive the spreader along with lime dropping out in little lines across the field. You did the whole field like this. If burnt lime was used, then you would shut the holes some, as you used less burnt lime than ground lime. Burnt lime came in paper bags, like cement, and was lighter than ground lime. If you exposed your skin, to burnt lime, it would irritate your skin causing sore places.

After spreading the lime, the next task was to disc the lime into the dirt. This would be the fourth pass over the field. Depending on the farmer and the machines he had, the next trip might be putting down fertilizer with the grain drill. If the farmer had a corn planter with a fertilizer attachment, then he would put down fertilizer when he planted the corn to save a trip across the field. We used the grain drill for the fertilizer, as Uncle John's corn planter did not have a fertilizer attachment.

The fertilizer came in paper bags, but before paper bags were common, it came in burlap bags, no plastic existed then. A burlap bag was made using heavy, small twine woven into a mesh and sewed into a bag shape. It was very heavy and durable material, and the bags had many uses on a farm. The empty burlap fertilizer bags were soaked in the branch or creek for a couple of weeks to get the fertilizer out of them. No one thought at the time that this was polluting the streams.

After the fertilizer was put down, a spring-tooth harrow was used for the final dirt preparation before planting. The spring-tooth harrow was usually pulled diagonally across the field so that the corn planter's marker would make a line in the dirt that was easier to see. This step would make it easier to get the corn rows where you wanted them to be.

Once the work with the harrow was complete, it would finally be time to plant. This was done usually in April. Some people relied on the whippoorwill to tell them when it was time to plant corn. A whippoorwill is a migratory bird that appears every spring. In the early evening after it gets dark, he will holler his call for several hours into the night. I never saw one in the daytime, and now I don't hear them much. The way you could get to see a whippoorwill was to listen to where the noise was coming from and approach the area in the dark with a turned-off flashlight. When you got close, what you judged to be within 100 feet, you turned the flashlight on and you could see the bird. He is about the size of a pigeon. I have a feeling with so many people now he has been pushed out of his habitat. This is a shame, as one more of God's creatures might become extinct.

We borrowed Uncle John's two-row International Harvester corn planter with its forty-inch row spacing to plant the corn. After most of the corn had come up, each row would be walked. When you spotted a missing hill, you replanted that hill by hand. You chopped in with a hoe, made a little sink area, dropped in a few grains, and then covered them up with soil. Imagine doing a ten-acre field like this!

After the corn was six to twelve inches tall, we had to go through the field and thin the corn. The object was to have only one or two plants grow out of each hill. If more were growing, you had to bend over and pull up by hand the selected plants and drop them in the middle of the row. This corn-thinning was a job I hated. It seemed to take forever and would get your back tired from bending over so much, and by this time the weather had gotten hot. A support stick helped with the bending, but that was about it. This thinning job took much longer than replanting. Both were total hand operations.

The corn had to be cultivated between the rows to keep grass and weeds down so that it didn't have any competition to impede its growth. It was cultivated two or three times until it got tall enough for the plants' leaves to meet in the middle of the row. When this stage was reached and the cultivation stopped, one would say he had "laid by his corn." The single-row cultivator was pulled by one animal walking between the rows. It had three narrow teeth staggered from front to rear and two handle controls to guide it with. Again, a lot of walking and time spent. As a result, some people employed a riding cultivator. This machine required two animals to pull it and straddled one row of corn. The driver would ride on a seat on the machine and only had to lift the teeth out of the ground at the end of a row.

When the corn had been laid by, usually sometime in June, it was on its own until it came time to cut it. Corn cutting time for the shucking corn was in September, usually the first part of the month. The weather could still be hot then. If the corn was to be put into a silo, it was cut by hand just like corn that was to be shucked for its grain. The stalks would be green and heavy. The arm-full bundles would have to be picked by hand and loaded crosswise on the hay rack wagons. This was very hard work. The wagons would be driven to the silo where a cutter was set up. This machine would cut up the stalk, ear and all, and blow it up a pipe that hooked over the top of the silo. It was powered by a flat belt attached to a tractor. The corn had to be pulled off the wagons by hand, several stalks at a time, and placed onto a moving feed table where the machine took them from there.

Corn-silage made good cow-feed. The corn was cut by hand, stalk by stalk, using a corn knife. This corn cutting knife had a handle about eighteen to twenty-four inches long, with a steel blade mounted on one end. The blade was set at an angle to the head, and the inside edge was very sharp. For a right-handed person to cut corn, you grabbed the stalk with your left hand about chest height and drew the knife back about six inches above the ground, and then struck. The stalk would go "whack" and be cut off with you holding it in your left hand. You would turn the stalk

Corn shocks of ripe corn set up in field. Picture courtesy of Sam Moore.

90-degrees and let it fall in front of you and then grab the next stalk, using the exact same process. You laid all the stalks the same way in a neat pile. When the pile got as big as an armful or as big as you cared to pick up, you started another pile. You would cut a small area like this with a bunch of neat piles lying on the ground. This would be several rows over into the field where you intended the corn shock row to be. After an area twenty feet by thirty feet was cut, it would be time to start a shock.

A shock consists of these cut armfuls of corn set up against each other in a circle. The cut-off ends of the stalks rested on the ground, and the stalks stood up just like when it was growing. The corn-shocks reminded one of Indian teepees. It would take four or five people working as a team, with each person grabbing an armful of corn stalks. They would then come together in a circle as a group, and each one would lean his armful against the other one by his side and carefully get the several armfuls to stand up. This may have to be done two or three times working together to get the shocks stable enough to stand up without someone holding on. After the small shock was stable, then they could work as individuals when picking up the armfuls and putting them on the shock, being careful to put equal amounts on all sides of the shock.

Shucking corn by hand out of a corn shock.
Courtesy of the Iowa Historical Society

If only one or two people were cutting the corn, then in order to start a shock required what is called a "horse." It was made with a small, long pole around sixteen feet with one end having two poles about five to six feet long attached together at the top of one point. The two short poles were put together in an A- shape and looked just like a giant letter "A" with the two longer sides and a short crossbar fixed between them to hold them in the A shape. The long twelve to sixteen foot pole was attached to the peak of the A with its other end resting on the ground. About three to four feet back from the peak, a one-inch diameter hole was drilled through the long pole. A broom-handle-sized stick about four feet long was stuck halfway through this in the long pole. This light device, called a "horse," was set up where you wished the shock to be built. One could lean his armful of corn against this long pole and the broom-handle-sized stick, resulting in the corn standing in place without anyone holding it. You had to be careful to place equal amounts of corn on all sides of these sticks. Once you had a small shock established that would be four to six feet in diameter, you had to reach in and pull the broom handle-like stick out. Next you had to catch hold of the giant A part of the horse and pull the long pole out from the standing stalks. If done correctly, you would have a freestanding shock and could continue to add armfuls of the corn to it.

After the shock was as large as you cared to make it, typically about ten feet in diameter at the bottom, a small twine string was put around the shock, at head-height and tied as tight as you could pull it. The shocks would be built in a row with about fifty feet or more between them. The next step was to disc and harrow the ground and plant small grain, either wheat, oats, or barley with a grain drill. It might be one month or more before the corn shucking was begun. This allowed the corn to dry out some.

The slowest operation of all was shucking the corn by hand. You grabbed several armfuls off the shock and threw them down on the ground. Then you got down on your knees beside the pile you had placed on the ground and began to shuck the corn. You had to pull the shucks off the sides of the ear of corn to where you could catch hold of the clean ear with one hand and hold the stalk at the base of the ear with the other hand. You gave a sideways snap to the ear, which caused it to break off from the stalk. The ear was tossed to an area where you would create a pile of ears lying on each other. You tossed the shucked stalks into a separate pile. The shucked corn stalks were called "fodder." After a good, sized pile of these shucked stalks had accumulated, they were tied up into a bundle with a twine string. The stalks were much lighter after the ears had been pulled off, and tying them together into a bundle made them easier to handle. These bundles would be fed to cattle, but it was poor quality feed compared to hay. The cattle would pick through them and eat what they wanted. What they left would rot away.

After the entire shock had been shucked, the fodder bundles were set back up into a shock again, and the pile of the corn ears was ready to be picked up. The fodder was hauled off the field on the hay-rack wagon which, of course, was loaded by hand. A box wagon approximately eight to ten feet long and three to four feet wide by two feet deep was pulled near the corn pile to haul off the shucked ears of corn. A basket called a "hamper" was used to measure the amount of the corn that was loaded into the wagon. The corn ears had to be picked up by hand and tossed into the hamper. As the corn was picked up, it would be sorted by tossing the good ears into the hamper. The short, partially filled, deformed ears were tossed aside and eventually were put into burlap bags and kept separate from the good ears. These deformed ears were called "nubbins." Every time a full hamper was dumped into the wagon box, a mark was made on the wagon box side. You would make four vertical marks, then for the fifth one would draw a diagonal line across the four vertical marks. I think this would be one barrel of corn. People always talked about how many barrels of corn were made. If you hired someone to shuck corn, they were paid by the barrel. The loaded wagon would be driven from the field to the corn house. Again, the corn would have to be thrown from the wagon into the corn bin by hand. A scoop shovel could be used in the wagon box, and this made the job go much faster.

The corn bin was constructed with wide cracks or spaces between the narrow boards. This allowed air to circulate and continue to dry the corn. By now the corn was finished unless you wanted shelled corn. We had a hand crank which we used to shell corn. You would put one ear after the other into it and could catch the shelled corn in a bucket.

Now fast-forward to the present time with no-till corn. The soil is not plowed or worked at all. A chemical herbicide is sprayed over the whole field to kill any existing or potential growing plants that would compete with the corn. Any lime applied is spread on top of the field by either truck spreaders or spreaders pulled with a tractor. Some of the fertilizer is applied the same way. The corn is planted by a planter, which is much heavier than a two-row horse drawn machine, along with fertilizer at the same time. No cultivation is done to the field. The corn is harvested by one machine, several rows at a time. This machine picks off the ears, then shells the corn as well. A truck hauls the shelled corn to a bin, where it is dried and stored. No hand has to touch the corn at any stage. In terms of manual labor, only tobacco required more hand work than corn. Thankfully, we didn't grow tobacco.

CHAPTER THIRTEEN

SEED CORN

When corn-planting season came, Daddy would go to the corn house to select good ears of corn to shell in the hand-cranked corn sheller. He used this corn in the corn planter for planting the next crop. Daddy always planted white corn. Most ears would be white, but every now and then you would encounter a red ear. This corn is known as "open pollinated corn." My oldest brother, W.O., acquired some yellow hybrid corn from T.W. Wood & Son Seed Company, in Richmond, Virignia. This was new to us, but we soon realized that the yellow hybrid corn would yield larger crops, everything else being equal, than the white corn. By planting only yellow hybrid corn, we could grow more corn on the same amount of land with the same amount of work.

Hybrid corn is made by taking the best features from different kinds of corn and actually "breeding" them in the field. The breeding is done by planting two rows of corn together that have a certain feature. Beside these two rows, you plant four rows of corn that have a different unique feature, but lack the unique feature of the first two rows. These sets of rows are then alternated throughout the field. As both sets of rows grow, they eventually reach the tasseling stage. It's very important that you go into the field at this stage and manually pull or cut the tassels out of the four-row set before they can produce any pollen. When the tassels in the first two rows fully develop (as these need to be left alone), the pollen from these two rows is carried by the wind onto the silks that develop where the ears are formed on the four rows of plants that were trimmed. As the ears develop on the four-row plants, they will have features that were previously only in the two-row plants, plus their original features. Under this cross-breeding approach, you have to be careful while harvesting the field to keep the four-row plants yield separate from the two-row yield. When corn from this four-row yield is planted the following year, it will retain the features from both sets of rows (the parent rows).

    Cross-breeding, while effective, is a slow process. Over the years, it has enabled the farmer to produce many more times the yield compared to what was produced over sixty years ago. As I remember, the white corn yielded forty to fifty bushels per acre. Today, farmers can expect on average one hundred twenty to one hundred fifty bushels per acre. Two hundred bushels per acre and higher are sometimes achieved. These hybrids, combined with advances in farm machinery, have enabled farms to grow much greater amounts of corn. However, the age-old law of supply and demand comes into effect, resulting in a lesser price per average. Even though farmers today can produce a much greater quantity and quality of corn, their bottom line has not significantly increased. On the contrary, it often times has decreased as the demand cannot keep pace with the supply. It takes a skilled manager to keep this from happening. I wish those luck that are trying to make a living by producing farm products.

CHAPTER FOURTEEN

Silage

In the winter, hay is the primary food source for horses. Since the 1940's, silage has become the primary winter food source for dairy cattle. This, I think, is due to advances in machinery that eliminated the hard manual labor required to produce silage. Silage is made by cutting up various plants such as corn, hay, or small grains into small bits or pieces. This chopped up material is very much like yard mulch in terms of its size and consistency. Silage has to be stored in a container to prevent its contact with the air. The material is cut up when the plants are green and contain high moisture content. Round tower silos were the first widespread storage containers for silage.

Early silos (which are now becoming obsolete) were small when compared to modern silos. Early silos were about twelve feet in diameter by forty feet tall or fourteen feet in diameter by forty-five feet tall. By the 1980's, silos had expanded to upwards of thirty feet in diameter by eighty feet tall as well as all sizes in between. These thirty-foot giants could hold ten or more times the smaller silos' capacity. When all the corn had to be cut and loaded by hand in the hot summer months, it would take a tremendous amount of effort to fill one of the smaller silos (refer to my chapter on corn).

Machinery that cut and chopped the corn in the field greatly reduced the amount of effort required to store silage. These machines were referred to as "choppers." The cut-up material is blown out of the chopper into a boxlike vehicle, typically a flatbed truck with sides six or more feet tall with an open top. It was driven beside the chopper in the field to catch the chopped feed as it was blown out of the chopper as it moved along. This was one way to get the corn from the field into the silo. Another collection and transportation method was the use of self-unloading wagons. Self-unloading wagons could be hooked to the chopper and used to collect the corn

Field chopper pulled and powered by farm tractor. Cutting corn in field and blowing it into an open top truck being driven along side.

as well as for transport. Regardless of what collection and transportation method was employed, the vehicle's contents had to be unloaded into the silo.

The chopped corn was normally unloaded and blown into the silo by using a blower. The blower was connected to a pipe that runs up the outside wall of the silo over the top. The blower would blow the chopped material up the tube and into the top of the silo. Early blowers had long feed tables

Corn chopper head on field chopper.

that were used to convey the material from the vehicle into the blower. In the case of the truck method, the truck was backed up to the blower feed table and the rear gate was opened. If the truck was equipped with a dump body, then it was raised to allow the feed to slide out of the back of the truck. Workers with pitch forks had to prod and try to control the material as it exited the truck and fell onto the feed table. If too much fell out of the body at once, it would cover the blower feed table. You could unload the

SILAGE

vehicle this way, but it still took some manual work to properly feed the blower. If the vehicle did not have a dumping mechanism, then the material had to be hand-forked onto the feed table.

Fourwheeled, open top bodied wagons pulled by a tractor were also used as hauling vehicles. A heavy canvas sheet was placed on the floor of the wagon prior to loading. One end of the sheet would be attached to a round roller on the rear of the wagon body floor. After the loaded wagon was moved into position behind the blower feed table and the gates were opened, the roller was slowly turned to wind up the canvas sheet. This caused the entire load to be moved to the back of the wagon. The materials were then hand-forked onto the blower feed table. A very similar hand crank device is now sold to unload pickup trucks. The wagon roller device was driven by a very slow-turning gear box which was powered by an electric motor.

/silage blower powered by a flat belt from a farm tractor blowing chopped up cow feed (like lawn mulch) through the metal pipe and up to the top of the silo where it falls down inside the walls. Feed stored in silos will be fed to the animals many months in the future.

Self-unloading wagons were developed that used endless floor chains and slats to move the material to the front of the wagon.

Unloading a farm wagon with chopped up corn into a blower set up at a silo. It was called "filling the silo." Courtesy of the Iowa Historical Society

They employed beater systems that knocked the material down into a cross conveyor that ran the feed out the side of the wagon. With these self-unloading wagons, there was no need for the long blower feed table. I don't believe you can even buy a long table silo filler blower today. Today, plastic bag storage systems have begun to replace storage silos.

With plastic bag storage systems, the feed is brought in from the field and fed into a machine that packs it into a long plastic bag about ten feet in diameter. These bags are white and, when laid on the ground, look like a long sausage tube from a distance. A tractor with a front end loader bucket or a skid-steer loader is used to dig out the silage for feeding from these plastic bags. This destroys the bags, requiring new bags to be purchased each year. Despite the costs of the bags, the quality of the silage stored in them is very good.

Really large farm operations now use cement bunkers to store their silage. The bunker is constructed by pouring a cement floor roughly fifty feed wide by one hundred feet long with walls roughly twelve to sixteen feet tall. The silage is then dumped from the field onto the floor and then picked up for storage by a tractor equipped with a loader. This packs the material tightly which is key in maintaining its quality. The whole top of the bunker pile is exposed to the air, but it can be and often is covered with a plastic cover. The trade-off to using these bunkers is the fact that there is typically some spoilage, but this is outweighed by the sheer amount of silage that can be stored this way versus the other systems (silo or plastic bag).

Over the past twenty-five years, a Total Mixed Ration machine or "TMR" has come into widespread use. A TMR is used to first mix and then feed the material to the animals. TMR's are typically mounted on a truck or have a set of wheels that enable them to be pulled by a tractor. Their bodies are mounted on sensors that scale the weight of the material loaded into them. A feed formula is calculated so that a certain amount in pounds of each feed ingredient is dumped into the machine. The scale allows for precise control of the various ingredients. TMR's have a slow-turning mechanical device consisting of either large augers, steel reel arms, mortar machine-type mixing arms, etc., that slowly rotate and mix the feed loaded into it. The machine is driven onto a feed bunk or feeding floor in a slow motion to feed the cows. This is quite a departure from feeding cattle from a five-gallon bucket like was done in our stable when I was young.

Farm tractor pulling a TMR (total mixed ration) machine, feeding animals as it is driven along the feeding trough.

CHAPTER FIFTEEN

HAY

The kinds of hay I first remember was a mix of timothy and clover that was cut in the last half of June, and lespedeza cut in the last of August or early September. All the hay cut was used for winter feed for the cows and horses. The June hay was the best hay for horses. From April through October, all the animals grazed the pastures. Very little hay was fed during these months. A horse-drawn mower with a five-foot cutter bar was used to mow the hay.

Soon after the hay is cut, it emits different smells. You may have heard of the saying "as fresh as newly mowed hay." It is a wonderful smell. On a farm you will encounter many smells or odors. Freshly plowed earth, hay, corn, wheat, oats, and barley all have their distinct odors. Manure, leaves in the fall, pigs, cows, horses, and chickens likewise have distinct odors. Some people say some of these odors stink. Whatever the smell, it's because God made that particular thing with its own signature odor.

Farm tractor with reciprocating cutter bar mower mounted on rear cutting hay. Courtesy of the Iowa Historical Society

After the hay is cut by the mower, it begins to dry out. It can't be safely stored in a barn until it dries down to fifteen or sixteen percent moisture. You learn to feel the hay to tell if it is ready to get up for storage. Hay put up too green or wet in a barn will get hot. In fact, it can get so hot it will set the building on fire. Many a barn has burned up this way.

The hay was raked up in windrows with a large two-wheeled dump rake. A windrow is a row or line of hay raked together. This rake was made with two large metal wheels on each end and pulled by a horse hooked up to the center of the rake. It had a series of large curved steel teeth spaced six to eight inches apart across its entire width between the two wheels. The rake had a seat in its middle above the teeth. The person driving the rake sat in this seat with the reins to the horse in his hands. To start raking the hay, you let the series of teeth down on the ground all at once with a hand lever, and as the rake was pulled forward, it gathered up all the cut material until it about filled up to the underside of the rake frame. This would be about two feet tall, as the rake teeth had a very large curve to them. When the rake had gotten full like this, you touched a pedal with your foot, which caused all the teeth to raise up quickly and dump the hay in a long pile. The teeth would return to the raking position very quickly after dumping the hay, so that all this was done without stopping the horse.

You drove across the field making these hay dumps, and when you got to the other side of the field, you would turn completely around and start back right next to the strip you had just raked. Next you would trip the rake at each place where you made the first pass. You kept doing this until the entire field had been raked. It was the first machine I ever operated.

For baling the hay in the field, a side delivery rake is much better than a dump rake. The side-delivery rake makes a continuous long windrow to one side as the rake is pulled across the field. It has a continuously

Raking hay with side delivery rake into windrow prior to baling with automatic tieing pickup hay baler.

turning reel made up of four or five bars with a series of teeth on each bar. These teeth swept the hay to one side to make a continuous row for the hay baler to follow later. It also can be used to flip a windrow over to let the underside dry out if it gets rained on before it gets baled. This windrow turning is tricky to do, as the rake has to be pulled with just the outer end of the rake touching the hay in windrow.

Next, a person would come along with a pitchfork and shock the hay before the field hay balers were used. A hay shock is about as tall as a person and about four or five feet in diameter. What this does is make the raked rows into a lot of separate piles of hay. Once the hay was placed in shocks, a wagon with a hay rack body is taken to the hay field and loaded by hand with a pitchfork. A hay rack body is about seven to eight feet wide and twelve to sixteen feet long, with each end of the body having a ladder-like frame about six or seven feet tall to pack the hay against. These wagon hay racks are also used to haul the corn stalk fodder or corn stalks for silage out of the corn fields. The hay racks were used to haul small grain bundles of wheat or oats from the field to the threshing machine. These were the three main uses for these wagon bodies, but they also had other miscellaneous uses on the farm.

It took one person on the wagon hay rack to pack the hay that the people working on the ground pitched up to him. He also had to drive the wagon from one shock to the next. The people working on the ground had to be careful when they picked up a shock of hay by sticking the fork deeply into the shock of hay to not get too much load and break the handle to the pitchfork. Sometimes a snake would get under these shocks. It got bad if the snake got pitched up on the wagon of hay. The person working on the wagon had to know what he was doing and use care to properly place and pack the hay by walking on it. Packing the hay was important so that it would not fall off the wagon beside the road before getting to the barn. If this happened, you would have to reload the wagon. Remember, this job was done in the summertime when it can be very hot and the dry hay loses small bits and pieces that gets all over you and sticks to you if you are sweating.

During this time period, some people had a hay loader, which was a separate machine that hooked to the rear of the hay rack wagon. To use this machine, no shocking was done. The hay wagon with the hay loader hooked on behind was driven over top of the length of the dump rake rows, or windrows if the hay that had been raked with a side delivery rake. The hay loader picked the hay up, and reciprocating arms pushed it over the rear hay post where it fell into the hay rack. It took one person to drive the wagon and two others to place and pack the hay that was pushed over the rear post. It also took time to unhook and hook up the hay loader between loads. While not perfect, a hay loader definitely lessened the amount of manual labor quite a bit. Unfortunately, we didn't have a hay loader.

The larger barns or stables usually had a hayfork to unload the hay wagon. This is why the front of the barn roof has a short peak sticking out about six to eight feet beyond the front wall of the barn. All barns have this peak

Barn with peak in front to support hay fork track.

profile that projects beyond the front wall of the building that makes it instantly recognizable as a barn. This was done so that when a wagon load of hay is pulled by the front wall of the barn, the hayfork would come out over the center of the load.

We had two hayforks. The older one was at the stable and had a harpoon-type hayfork. This type of hayfork had two slender steel prongs about three feet long, each that pushed down in the loose hay on the

DICK'S BOOK ~ REFLECTIONS OF A FARMER

wagon. These two steel prongs had a groove in them that allowed a small rod to be recessed into the groove. This rod worked a finger-sized steel part that pivoted near the end of the prong to project out ninety degrees from the prong. This kept the hay from slipping off the fork. These two small fingers had to be set for each load of hay it picked up. These fingers are released by pulling on a rope after each load was in position up in the barn loft. A very long three-fourths inch to one inch manila twine rope run over pulleys was used to lift the hayfork.

Grapple type hay fork. Fork attaches to lift rope and pulley from carrier under roof by ring on top of fork.

A steel track was fixed to the underside of the barn roof rafters. This track ran the entire length of the barn roof, including the peak that sticks out in front of the barn. A four-wheel trolley rolled up and down this track. The four-wheel trolley was called a "hayfork carriage." A catch was placed on the track at the end of the front peak to hold the carriage over the center of the load of hay. A long hay rope was anchored to the front of the carriage. It went down to and around the pulley on top of the fork and then back up and over a pulley on the rear of the carriage. This was called a double-line lift. The rope continued just under and parallel to the track to the rear of the barn, and over a pulley fixed to the underside of the roof. It came off this pulley and ran down the outside of the back wall of the barn to a similar pulley anchored near the bottom of the barn. The rope then extended from the bottom pulley at the rear of the barn to where it was hooked to a horse.

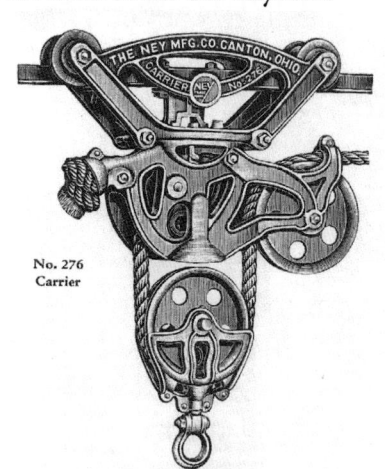

Hay fork carrier on track that is fixed to underside of barn roof rafters. The hay fork attaches to the ring under the bottom pulley.

Hay fork lifting its load in mid-air between wagon and barn roof. Courtesy of the Iowa Historical Society

With the fork set in the hay on the wagon, the horse was driven away from the barn, pulling the hay rope. With the double rope from the fork to the carriage, for every two feet the horse walked, he only lifted the load one foot up in the air. This makes it easier on the horse. When the pulley on the hayfork struck the carriage, it goes into a pocket that held it and also released the catch on the carriage so that the carriage with the hayfork load of hay rolled back into the barn. When the carriage got near the place where the person working in the barn loft wished the hay to be dumped, he had to holler or call out or signal to the one driving the horse to stop. The person working in the loft had to then catch hold of the trip rope that was dangling in the air and pull on this rope to release the two fingers on the fork prongs so that the hay would fall off the fork. The trip rope is a small rope just long enough to reach from the roof to near the floor. It was tied only to the fork trip mechanism. He also had to be careful to stand to one side; otherwise the fork load of hay would come down on top of him. He then had to take a pitchfork and move the hay to the sides of the building and neatly keep it packed up. It got very hot in a barn loft in the summer, as you didn't get any breeze or air stirring. It was much worse than working outside which is bad enough.

Now, the carriage and fork had to be pulled back to the front of the barn for the next load to be lifted. This was done by the pullback rope, which is another smaller rope long enough to reach from the wagon in front of the barn to the rear of the barn. The person on the wagon had to pull on this pullback rope to get the carriage back and locked in the catch on the track at the front of the barn. Once the carriage locked in the catch, it released the pick-up pulley on the fork, and the weight of the fork let the fork fall back to the wagon of hay. The wagon person sat the fork into the hay and locked the trips, then hollered to the person driving the horse to pick up the load. This was repeated as many times as it took to unload the wagon. On the last lift, the wagon person had to be careful not to catch the fork on or in the wagon floor. He had to check the trip rope and pullback rope before each lift to see that they didn't catch on some part of the wagon or barn and cause trouble.

If the hay was to be put in a shed or some place that did not have a hayfork track, then the load would have to be unloaded with pitchforks by hand. One person worked on the wagon to fork the hay to one or more people that would be working in the shed lofts packing it back into the building. In some cases, the hay was stored outdoors in ricks or stacks. A pole stack was made by setting up a long small post about sixteen to twenty feet long. One end of this pole was stuck in the ground, like a very tall fence post. The hay was packed around the pole and the stack would be built up into the air around this pole. A hay stack was made round like a silo. The pole was used to give stability to the stack, so that it would not fall over after it was stacked so high. The top of the stack would be rounded off so that rain would be deflected to the outside. A hay rick was bigger than a hay stack and when finished would look like a giant loaf of bread. The rick would be long with a rounded top to make the rain water run off to the outside.

The hay stacks or ricks were usually used up during the first part of the winter. The longer the hay is left outside exposed to the weather, the quicker it becomes spoiled and unfit feed for the animals. To make hay much easier to handle and transport, hay presses or bales were made. These were packed hay in a rectangular shape about sixteen to eighteen inches in height, twenty-four inches in width and about thirty-six to forty inches long. They typically weighed anywhere from forty to one hundred fifty pounds, depending on the size of the bale and how tight it was packed.

    I remember the first bales I ever saw. I was coming home from school one afternoon. Daddy had gotten someone with a horse-powered baler to bale a stack of hay. The horses walked around and around in a circle to make power to turn the baler. A hay baler worked by using a plunger to push the loose hay into a long chamber. The size of the chamber determined how high and wide the bale would be. The length of the bale was determined on the horse-powered baler by the person feeding the baler with a pitchfork. To start a bale on these types of balers (sometimes called "hay presses"), a thick three to four-inch wooden block almost the same size as the bale chamber was put into the chamber. Hay was packed against this wooden block by the plunger which strokes back and forth by a crank arm just like a piston does inside a motor. As the hay is continually pushed back into the bale chamber, it got tight or packed because it resisted sliding through the chamber. It took power to push this hay back into the chamber. After the block went back a set distance being pushed by the hay, say about three feet, it struck a bell which told the person feeding the baler it was time to drop in another block, which he did. If you go to one of the farm antique shows in the summertime, you might see one of these balers actually at work baling hay or straw.

    The person running the hay baler then continued to feed in more hay, which pushed the front blocks further down the length of chamber. These

Stationary hay baler that has to be fed by hand with a pitchfork. Baler is powered by a flat belt hooked to a farm tractor.

wooden blocks had slots or grooves on both sides of their faces. A long precut small-diameter wire was stuck through these slots so that the wire completely encircled both sides and ends of the bale. The wires had to be tied before the bale got to the end of the baler. Some hay balers used three wires. All I ever saw were ones that used two wires.

As the wooden blocks fell out the back end of the baler, they had to be carried back to the shelf where the person feeding the baler could get them to put back in when the signal came. The bales of hay had to be packed away from the back end of the baler. To handle wire-tied hay bales that might weigh one hundred pounds or more, you used hay hooks. The small wires would cut your fingers if you try to lift the bale by its wire.

With a horse-powered baler, each bale was weighed on a platform scale. A small wooden stick, like a popsicle stick, was stuck between a wire and the hay. The weight of the bale was recorded on this stick. It was difficult to get these small flat sticks between the wire and the hay on the end of each bale. By weighing each bale, it could be determined exactly how much hay was baled for that customer. The person with the baler charged for baling by the ton. The farmer sold his hay by the ton, as well. It took a crew of five or more people to run one baler.

J. L. Case hay baler with 2 people riding on each side of the bale changer. One person sticks the wires between the bales and the other one ties the wires. Courtesy of the Iowa Historical Society.

A few years later, one could get a baler that baled the hay in the field right out the windrow, which saved all the pitchfork work of getting up the hay. The first field hay baler I saw was a Case wire baler. It required a person to drive a tractor to pull the machine. It required two people riding on each side of the baler chamber.

One person pushed the wire through, and the person on the other side tied the wires. Talk about a dirty job, this was one of them. Each time the plunger hit the dry hay, it caused dust and chaff to fly out of the machine, which soon got all over a person riding on the machine. Some would tie a large cloth over their face to help from breathing so much of this dust. They would then look like a person about to rob somebody. I don't think the little mouth and nose filters we have now existed then.

In the late 1930s, a person by the name of Ed Nolt in Lancaster County, Pennsylvania, made a twine baler that tied it self automatically. No person was required to ride on this field hay baler. All it required was someone to drive the tractor. A group of four investors that had purchased the new Holland Machine Company in New Holland, Pennsylvania, (which is in Lancaster County), made a deal with Mr. Nolt to make and sell his baler. This New Holland baler became famous very quickly, as it changed the way farmers got up their hay.

1950s era square hay baler. Makes bales weighing 50 - 80 pounds or less.

We got one of these New Holland #76 hay balers in the fall of 1945. Not only did we start baling all our hay and straw in the field, but also did it for the neighbors for a price, I think, of twenty-five cents per bale. Think of all the hand pitchfork time it saved and all the steps that the pitchfork required and

Loading bales of hay onto flat bed wagon.

how little hay you got up. Of course, the bales had to be picked up out of the field, but a crew of four people could get up many times the amount of hay in bales, compared to getting the hay up loose. This work was cut down even more if you hooked a wagon behind the baler and loaded the wagon right out of the bale chamber without letting the bales drop to the ground. This would require one or two people to ride the wagon to pack up the bales on the wagon.

After we got our own hay baler, we started growing alfalfa as our main hay crop. Alfalfa is a superior plant in that it has a lot of food value, especially for dairy cows, and can be cut four or five times a year. Alfalfa produced more tons with multiple cuttings than any other hay. It soon became our main hay crop. After we added onto the dairy barn in 1951-52, we installed a hay dryer in one section of the hay loft. This hay dryer used a large electric motor-driven fan that forced air into a duct. The duct distributed the air out onto the floor of the loft. A second floor was built on top of the main floor about six inches above the main floor. The second floor was slotted to allow the air to come up into the hay bales packed on top of it.

Hay baler with hay wagon hooked on behind to catch bales directly out of the baler. Saves having to pick them up off the ground. Courtesy of the Iowa Historical Society.

With this dryer, you could bale the hay sooner because it did not have to be completely dry before you started. Sometimes you could beat a rain coming up and not let the hay get wet. The dryer we had could hold 500 bales. Sometimes it took a week of continuous air blowing into the hay to get it completely dry. If the weather turned rainy after you loaded the dryer, you still had to run the fan to keep the hay from getting hot. When the natural air turned drier, then the hay would dry much quicker.

After the hay had dried and was safe to pack back in the barn, it had to be handled again with the hay fork system to unload the dryer and get it ready for another cutting. This was a lot of work, but it produced some really good cow feed for the cows in winter when they had no pasture to eat.

1970s era round hay baler with round bale. Makes a bale weighing 1,000 pounds or more.

After a few years it seemed that nobody liked to handle all these bales, so they made balers to make a giant round bale weighing anywhere from 1,000 to 2,000 pounds which could be handled by a tractor with a front-end loader. Now no one had to touch the hay to get it up. In the same fashion now, no one has to touch the wood to get the trees from the woods to then be sawed into lumber. It is all handled by machines.

The objective of growing and storing hay was to have something to feed the animals in the winter months when snow or ice may be on the land. Grass and hay does not grow much at this latitude (38-40 degrees) from November to March. In order for these domesticated animals to survive during these winter months, you have to cut and put in storage the plants they eat. This food has to be fed to them during the time they cannot find enough to eat out in the fields. The hay that was put in the barn or the stable lofts was forked to them daily with a pitch fork. When the loft is filled with hay in the summer, a chute or hole is left in the hay, which was usually formed by a upright frame from the loft floor to near the roof. The animals had stalls or an area on the ground floor of the barn with hay stored overhead in the lofts. One had to climb up to the top of the hay by a ladder built on the side of the hay chute in order to take a pitchfork and throw down the feed for the animals.

After the hay had been thrown out of the loft, it had to be distributed to the stalls or put in the mangers for the animals to eat. Again, more pitchfork work. This is how the hay was gotten out of the barn that had been put in by the hay fork. A little was taken out each day. For feeding out of the hay stacks or ricks, they were usually built on the edge of the field the animals were kept in. A temporary manger and fencing would be set up right beside the stack or rick, and each day some of the hay would be forked to the animals just like it was in the barn. This was for animals that had no shelter in these fields. A barn or shed is much better for the animals especially if it rained or snowed, in that they can keep dry. If they are dry, the cold does not bother them as much. When a bunch of cows are in an enclosed barn, they will make enough heat from their bodies to make the whole enclosed area comfortable.

For feeding the bales that were stored in the same place as the loose hay, again one had to climb up to the top of the hay, pull out each bale, and throw it down the chute. It was easier to keep up with the quantities this way, as you could keep a count of how many bales were handled. You can handle bales faster than loose hay with a pitchfork. For feeding bales to animals out in the field, a number of bales were put on a flat bed or hay rack wagon and carried out to the field where they would be thrown on the ground with some distance between each bale. This would allow the animals to be scattered out, so that they did not trample on the hay so much. By feeding in one area of the field one day and going to a different area of the field on another day, you would do very little damage to the field. If you feed in one area all the time, it soon becomes all mucked up, and whatever is growing on it becomes ruined. This is one disadvantage to feeding big bales in a feeder set in one spot. Machines are available now that allow a big bale to be spot-fed like the small ones are done.

CHAPTER SIXTEEN

WHEAT AND OATS

In addition to corn, the corn land was planted at intervals with wheat, oats, rye (for the cows to graze), or barley that would be threshed for grain. All of these small grain crops were planted in the fall. Grass seed would be planted along with the grains so that after the small grain was harvested, a crop of hay would take its place which would be cut later in the summer months. These hay crops were cut every year for two or three years. Thus we had a rotation of corn, small grain, and hay on land that would, after a couple of years, be turned into pastureland for the animals to graze. With this type of crop rotation, the land would be plowed only once every five to six years.

The soil preparation for small grain was performed with a disc followed by a harrow, with the grain and grass seed sown by a grain drill. The grain drill would put the grain seeds in the earth in rows that were seven inches apart. The grain would ripen in June the following year. You can tell when the grain is ripe and needs to be cut or threshed when

Grain binder in transport or road mode. Note cutter bar and wide canvas belts and reel. Essential components for the machine to operate.

you see the grain changing color from green to yellow or when the husk opens or becomes loose on the heads of the grain. When the grain was cut with the binder, you can cut it sooner than if it was to be cut and threshed with a combine. Before binders were made, the grain would be hand-cut

DICK'S BOOK ~ REFLECTIONS OF A FARMER

Bull wheel on ground that powers the machine as it is pulled forward. Machine is now in field mode.

with a cradle. When I was growing up, binders had been in use for many years. The binder was towed into the field lengthwise like a trailer on two wheels. Once the machine was in the field, the two wheels were removed and typically placed by a fence. The main drive wheel, called the "bullwheel," was cranked down onto the ground by a built-in jack system. This bullwheel was used to power the binder. The tongue was transferred to the front of the machine and a small smooth wheel on the outer end of the cutter head table supported that end of the machine. The majority of the weight of the machine was supported by the bullwheel which had cleats on its outer rim to allow it to get traction to drive all of its functions.

A team of two or three animals would be hooked to one side of the front of the machine to provide power. Once up and running, the binder used a reciprocating cutter blade to cut the stalks of the grain about six inches off the ground. These stalks would be knocked or pushed by a large six-foot diameter slow-turning reel, backwards onto a wide continuously moving canvas belt. Attached to the canvas belt were small wooden slats which moved the stalks to one side. This canvas belt ran in a flat plane. It trans-

Grain binder in operation pulled by three horses. Note round twine box on rear of machine.

ferred the stalks to two additional, continuous moving canvas belts with small wooden slats of their own. These two canvases were set up at an approximate forty-five-degree angle. One formed a bottom floor and the other rested on top as a ceiling. They were around two inches apart. The top canvas turned in unison with the bottom one in opposite directions, so that the grain stalks that were in between the two conveyor belts would not tumble back down the forty-five-degree incline. These two opposite turning canvas belts deposited the grain stalks onto a downwardly sloping table where moving finger-like bars packed the stalks into a bundle about eight to twelve inches in diameter.

Grain binder in operation cutting ripe wheat and putting it into tied up bundles.

When the grain stalk bundles reached a certain size, a long needle was tripped that carried a twine string around the bundle to automatically fasten it. (The tied-up grain stalk bundle was referred to as a "sheave," when it was cut by hand with a cradle.) The bundles were collected into a bundle carrier. After collecting about a dozen bundles, the driver triggered a foot pedal that caused the bundles to fall from the carrier to the ground. The binder basically cut the grain, and

Taking a bundle of grain dropped by the grain binder and flattening it somewhat to make a cap for the bundles that are stood upright. This is done to make a sort of roof for the shock of grain bundles to deflect rain to the outside of the shock.

gathered it into small round bundles tied in string, in one automatic continuous process. The binder was a marvel of engineering as it evolved from McCormick's original reaper which could only be used to cut the grain. By comparison, when using a reaper, another person had to walk beside the machine and use a long handle rake to manually pull the stalks off the table into a pile onto the ground. The pile then had to be gathered and tied by hand. Even though the reaper was a vast improvement over a cradle, it was still very crude when compared to the binder. After many years, binders evolved to where all the operations on the grain were performed automatically by mechanical parts arranged in an ingenious way.

With the pile of grain bundles scattered over the field, workers had to come along and set them up into what we referred to as "shocks." The bottom part of the shock is formed by just standing the grain bundles with the cut stem ends on the ground in a circle approximately four feet in diameter, like making a corn shock. Several additional bundles are then laid on top of these upright bundles and other bundles are flattened and fanned out forming a roof like structure.

Grain shocks set up and scattered about in the field after it has been cut with the binder. Courtesy of the Iowa Historical Society

The bundles were arraigned in this fashion so that any rain will run off to the outside of the shock, thus keeping most of the grain dry. When grain was cut with a binder, it still needed to dry out more before it was ready to be threshed. Depending on the weather conditions, these shocks would be ready for threshing after about a month. As a result of using a binder, the only manual work in the operation was in assembling the shocks. Through research, you can find pictures of these grain shocks scattered in fields. They make a pretty picture.

WHEAT AND OATS

In order to get the grain into grain bins in a building, it had to be threshed. Threshing is the process that separates the small kernels of grain from the stalks. This was accomplished through a large box-shaped machine called a threshing machine. The threshing machines we used required a team of about twenty people to operate. The first step was to select a location to set up the machine, which for us was typically in the lot near the stable. We selected this location as we needed a place for the resulting pile of straw (called a "rick"). Other times, we would set up the machine in the field where the shocks were. I remember once Uncle John had some grain that was in the back field of Cherry Grove, and they set up the machine right there. (This particular location is where my nephew, Samuel Nelson Harris', house resides today.) A thunderstorm came up very quickly while they were threshing, and as there were no buildings to take shelter in, the workers had to seek cover under the wagons. It was very dangerous, as lightening strikes were close by which frightened the horses that were still hitched to the wagons. The workers that were driving the wagons had to stay out, unsheltered, to hold onto the reins to try to keep their teams under control. When a team of horses are frightened, they have a tendency to flee the situation, which often destroys whatever they are attached to. It's a real challenge to control a team of frightened horses once they start to get out of control. Thankfully, I was never involved with a runaway team of horses, but I heard stories of them. Needless to say, there was no more threshing that day as the grain needed to dry.

The threshing team that worked our neighborhood farms was owned by Mr. Joe Wingfield. He owned a John Deere "D" tractor with steel wheels that moved slowly over the dirt roads, rattling the driver all the time. The tractor was used to pull the all-metal threshing machine that had a large, long pipe that blew the straw into a big rick. The workers would dig holes in the ground for the wheels of the threshing machine to rest in. They had to be very careful to keep the machine level, thus some holes had to be deeper than others to compensate for the grade of the land. They would

Threshing machine powered by flat belt driven by farm tractor (not shown). Loaded wagon from the field with the grain shocks pulled along side of the feeder where the bundler are pitched by hand into the machine. The straw is blown out the rear of the machine through the large pipe that can be moved up and down and side to side to make a large stack. The threshed grain is put into the wagon box parked on the right of the machine. Courtesy of the Iowa Historical Society.

even go to the extent to check the level of the machine with a carpenter's level. Once level, the machine was anchored to the ground with stakes so that once the machine was connected to the tractor via a long, flat belt, it would not move.

It took five or six people to work at the threshing machine itself. Two or more worked at the bagger, which is where grain exited the machine and was put in burlap bags. One or more had to then load these bags onto a wagon which were then taken to the grain bin and dumped. At least two people had to pitch the bundles into the threshing machine. The wagon driver could assist with this. One person had to operate the straw pipe to build the straw rick, which was the easiest of the jobs. The steam engine-powered threshers took even more people to fire the boiler and to transport water for it. Five or six wagons were needed to haul the shocks from the field to the threshing machine. Five or six additional people with pitchforks were needed to load the bundles onto these wagons. Mr. Wingfield would remain with the machine to observer and adjust it as necessary. Since most farms could only supply three or four workers, neighbors would pitch in to help one another to make up the rest of the work crew. When they finished at your farm, it was the unspoken rule that you went and helped the next neighbor with their field. No compensation between neighbors was ever passed for this labor.

One of my main memories of the threshing process was of the dinner, or noon meal. My mother would arrange for help on the day our fields were threshed and set up two tables to feed the work crew. One table would be on the back porch or under a tree in the yard, and the other was the regular dining room table. Imagine having to cook and prepare for about four times as many people as was done at the normal meal. That's what Mother had to do. No one wanted to earn the reputation of serving a skimpy meal, so you can imagine the feast that was prepared. Typically a ham would be cooked ahead of time, plus chickens would be killed and prepared to serve as the meats for the meal. As the threshing would typically take place in July, she would supplement the meal with many fresh vegetables. Desserts would normally be cakes and pies of different varieties. This might be why these meals are my most-remembered thing about the threshing process. Sometimes young men met their future wife at these threshing machine dinners.

The invention of the combine ended the threshing machine era. The combine was the, result of the combination of the binder and the threshing machine. It combined the components of both machines and thus scaled down the entire process. The straw was ejected from the rear of the machine back onto the field. If desired, it could then be raked and picked up with a baler (which we did). The combine only required two workers if the grain was put into burlap bags. If the combine was set up to handle the grain in bulk with a grain tank, it only required one person to drive the tractor pulling the combine. When the grain was bagged, a second worker had to ride the combine to bag the grain. This was a dirty job, as the machine produces a lot of dust and chaff that would cover the person performing this task. When the yield was good, the bagger had to work hard to catch a full bag, tie it off with a "miller's knot," and hook an empty bag back onto the machine. A gate was used to swap the nearly full bag to an empty one with both bags sitting side-by-side. Sometimes the bagger would have to get the driver to slow down in order to not fall behind. If you let the bag get so full that it couldn't be tied, you would have to make the driver stop to remedy the situation. The bags of grain would be put into a chute that held three or four bags at a time. When the chute was full, the bagger would pull a rope that would let the bags slide to the ground. Later you would load these bags by hand onto a truck or wagon to be taken to the grain bins and unloaded. Even though the combine greatly reduced and simplified the threshing process, it was still hard, tiring work.

Combines with grain tanks pulled by farm tractors. Courtesy of the Iowa Historical Society

The combine allowed the farmer the luxury of being able to thresh his grain at his convenience. No longer would he have to wait for the threshing crew; rather, he could thresh at his own pace. There were no scheduling problems, and the wife did not have to prepare a huge meal. Many of the farm antique shows feature threshing machines in action; thus you can observe how the machine worked. At these shows, you won't get the same home-cooked meal that the threshing machine crew was famous for, but you can buy similar foods at the show. Just as with other farm crops, the amount of manual labor was greatly reduced through advances in machinery. One person with a combine could produce many times more the output with the same amount of work. Today, the advent of large, self-propelled combines with thirty-foot-wide cutter heads have brought the threshing part of the machine back to a similar size as the old stationary machines. Another significant improvement is that the operator can work from the luxury of a climate-controlled cab, thus eliminating having to deal with the dust, chaff, and the heat. Costs for these machines start at $200K and up depending upon the accessories you choose, just like choosing accessories on a car. You can even add stereo and GPS guidance where you don't even have to steer. There's always something bigger and better on the horizon. Where do these new features end? They never will.

CHAPTER SEVENTEEN

CHICKENS AND TURKEYS

My mother always kept chickens and turkeys. In fact, she ran the household with the income from them. She had the laying hens in two chicken houses from which the eggs had to be gathered every day. After the eggs were gathered, they had to be washed, wiped dry, and packed in the egg cases. One egg case contained thirty dozen eggs. One market or outlet she had for these eggs, turkeys, and chickens was a gentleman by the name of Walter Swift. Mr. Swift lived near Mine Run, Virginia, and traveled the countryside in his truck buying eggs and poultry from the farms. He would take these food items to Washington, DC, and sell them direct to the big-city customers. He did this every week. In addition, in the town of Louisa, a Mr. Snyder bought our eggs, live chickens, and turkeys. However, Mr. Snyder did not come to our house to purchase them the way Mr. Swift did. I think there was some negotiation on how much she got paid for the eggs and poultry. A small child is not much concerned about money, so I don't remember much about this, except who got the eggs, chickens, and turkeys. Sometimes she would sell butter this way as well.

During these years there were many markets for products that the family farm produced. There were livestock markets for live cattle, hogs, and sheep. There were markets for milk and cream from the farm. There were markets for live chickens, turkeys, and eggs. There were markets for corn, wheat, oats, and hay. Almost every town had one or more of these markets. Today, a farmer has very few options as to where he can sell his produce. Now a farm specializes in just one product which may have a market more than a hundred miles or several hundred miles away. Thus has agriculture changed in my lifetime.

CHICKENS AND TURKEYS

Guess who had the job of feeding, watering, getting up the eggs, and sometimes cleaning the chicken house? Yours truly. Actually, I helped my mother do these things. Every spring, my mother would get little baby chickens from Hall's Poultry Farm near Mineral. These little baby chickens would be put in the brooder house where there was a device called a "brooder" (hence its name). The brooder would keep the chickens warm. The brooder house had to be cleaned out, disinfected, and fresh wood shavings put down prior to the arrival of the baby chickens. Chickens were fed what is called "mash." Mash is a finely ground-up shelled corn and other small grains ground up along with minerals so that the chickens have a balanced ration. We purchased our mash from the Mineral Milling Company, which was located where the present post office is located in Mineral today.

The Mineral Milling Company was powered by a large one-cylinder diesel engine that went "bong, bong, bong, bong" and could be heard all over town when it was running. Sometimes Mother purchased the chicken feed from the Louisa Purina Store. A lot of the times this mash or chicken feed came in cotton bags. When these bags were empty, they were washed and opened up and used as material for making dresses, shirts, bed sheets, or other apparel for a person to wear. If the bags had printing on them, it could be bleached to make the material totally white.

The baby chickens would grow up to frying-size at which time some would be killed and we would have fresh fried chicken. KFC or any of the other food chains do not have chicken that equals the taste of these fresh fried chickens. What we didn't eat would be sold as fryers. The females would be allowed to become full grown and were called hens. They would begin to lay eggs in the fall and be kept until they were two or three years old. Afterwards, they would be sold as "old hens."

To me, the turkeys were the most work. Mother would always have two or three turkey hens as a well as one male "tom" turkey. The hens would lay their eggs in early spring, late February, or all of March. Even though our brown turkeys were supposed to be tame, they had a lot of wild ways. A turkey hen is very secretive about where she makes her nest to lay her eggs. I had to watch the hens and be very careful to keep well back and a low profile to see where they went to lay their eggs. I was kind of like a detective tailing a person to see where he was going. This was no easy job, as the hens were very clever as to where they made their nests. You could get within a few feet of a nest and not see it. We needed to be able to go to the nests daily and get the eggs one by one as they were laid. The reason for gathering the eggs daily was that raccoons, weasels, foxes, and other wild animals would find the nests and eat the eggs. These animals loved eggs and sometimes got in the chicken houses which were near our house and ate the hen eggs or the chickens themselves. Usually we ended up taking one egg per day. One egg was always left in the nest; otherwise, the hen would get wise to someone stealing her eggs and abandon her nest. Some days they didn't lay any eggs.

You must think I lived in the wilderness. Actually, the country was fairly open, with about one-half of it woods where we lived. Some people had more trouble than others with wild animals. One wild animal that was rare were deer; they were rarely seen. Not any more.

One turkey egg-gathering incident I remember started normally. I went to the turkey nest and gathered the new egg I found there. On the way back to the house, I carelessly dropped it and broke it. I continued on towards the house and went in. Upon entering, mother asked me where the turkey egg was. I said, "she didn't lay." This was a lie, of course, but it could have been true as some days the turkeys did not lay. A day or two later my brothers discovered the broken turkey egg. They reported to my mother what they had found. She confronted me as to what actually happened and said how disappointed and ashamed she was of me. No child likes to hear this, but when one does wrong they need to be dealt with. I cannot remember the punishment, but I do remember the words of how I had hurt her. This hurt me more than a switch ever could.

The gathered turkey eggs were kept together and after the hens quit laying, the eggs were put under a setting hen. In order for any eggs to hatch, first the eggs must be fertile. That means the male must have inseminated the female. The hen bird lays the eggs and when they have a group of eggs, the hen will do what is known as "set." What the hen does is actually set on the nest of eggs twenty-four/seven. This keeps the eggs warm. The hen will not get off the nest. If you force a setting hen off her nest, she will raise a big fuss and potentially peck you. This setting keeps the eggs warm, and after two or three weeks, the baby birds will hatch or come out of its shell. The time span from when the setting begins until the baby bird hatches is called "incubation." Man has learned to duplicate this setting process with electric incubators. The eggs are put into the incubator and kept at a certain temperature and are turned over periodically, just as the hen turns over her eggs. After a certain time span, the little baby chicks emerge from their shells.

What I have described is a miracle. How can a pile of half-liquid material change itself into a living creature that has life? Who understands it? I say only God or a very highly intelligent being could conceive of such a plan, then create it and start such a life cycle of an egg becoming a bird and that bird producing more eggs and so on. Now, which did he make first, the egg or the bird? How did any life first get its start? Certainly life changes over very long periods of time. But how did it first start? Some say there was a "big bang" and life was started. Who made the "big bang"? Did it happen without a plan? When we build something, whether it is a house or a machine, you first have to have materials to work with. All these materials, earth, air, water, etc.,-who made or created them? In the Bible, God asked Job, "Where were you when I laid the foundation of the earth?" You can deny the existence of a higher being if you want, but that does not change the fact of its existence. This life cycle which we are all a part of is most dramatically presented on a farm such as I had the good fortune to be born into. What we do have is a choice of whether we will beget children or not, what they will be taught, what kind of role model we will be to them, and whether we will acknowledge that there is a higher being and worship Him. A lot of people never go to church. How sad that they are not willing to acknowledge and worship the ONE that put this thing we call life into existence.

Now, back to the baby turkeys that have just hatched. Mother uses little coops about the size of a dog house that were placed around the farmstead. These coops had roofs to keep the rain off the birds. A mother hen and some of the baby turkeys were put into each of these coops. At this stage in a turkey's life, the mother hen is very protective of her babies. Feed and water had to be taken to these coops daily and the birds were let out of their coops during the day. The coops were built with a door so that the birds could be shut up at night. This was to keep the wild animals from getting some of them at night. About dark, someone had to make sure that all the birds were in their coops and then shut them up. After the birds were about half grown and after they had learned to fly, they would begin to roost (sleep) in the nearby trees. They would begin to forage for food, mostly insects, out in the fields.

Our nearest neighbor was Uncle John and Aunt Lucy. It was about one-fourth a mile between the farmsteads. They had turkeys too, and our turkeys would go down to Cherry Grove (the name of their farm) and get mixed up with their turkeys. I think our turkeys were marked by clipping some tail feathers. I know we had a way of telling whose turkeys were whose and, also, had a number count on how many turkeys we had. When the turkeys were all together, they would all look alike. Mother wanted her turkeys home every night. Every fall from sometime in September until the

CHICKENS AND TURKEYS

turkeys were sold (either before Thanksgiving or Christmas), the turkeys had to be separated and driven home. This was a thankless job.

Every afternoon after school when the sun had gotten low, but before the turkeys went to roost, mother and I would go down to Cherry Grove and separate our turkeys out and drive them home. This may sound simple, but sometimes it was quite a hassle, because the turkeys didn't want to be separated. Their instincts were to be in a flock. To perform this job, you needed a light stick about four or five feet long. Walking in among the group of turkeys, you could use the stick to prod a bird to one side and gradually get your birds to one side of the group. There would be much running back and forth and using your stick as well as shouting at the birds. Finally you would have your flock separated out with the correct count. You could make the birds walk along ahead of you, which was not as difficult as getting them separated. What would happen more often than not was we would get the flock up to the area where the present silos are (about one third of the way home) and then some of the turkeys would decide to fly up in the air over our heads and go back to Cherry Grove. Sometimes I cried when this happened. We would then have to go back and perform the separation process all over again. We had to get them through two barbed wire fences and after we got them through the second fence (at what was called the "walnut tree gate") they would sometimes run ahead of us on home. Boy, was I glad when those turkeys got sold!

CHAPTER EIGHTEEN

THE SAWMILL, LOGGING AND THE FORESTS

    In the summer months of July through August in 1936, my father had a person by the name of Allen bring a sawmill and crew to our farm. They cut about ten acres of pine trees, converted it to sawed boards (i.e. lumber), and hauled it away. This outfit was referred to as "Allen's Sawmill." The mill was set up near the center of the tract that was to be cut, which in our case was at the bottom of a little hill on a slight downhill incline. The incline made it easy to roll a four-wheel rail cart down a track that was set up to move the lumber away from the saw. As the logs were sawed up, the lumber was flipped over onto the cart. The slabs or outside planks which were cut off the logs were carried by hand across the track and dumped in what became a very large pile. Slab planks were mostly used for firewood, but could be used for the construction of outside structures as well. When the cart had a partial load, it was rolled down the incline track where the various sizes and lengths of boards or lumber were stacked in neat, separate piles. When these stacks were large enough, they were hand-loaded plank by plank onto a flatbed truck. My brother Fred told me that the lumber went to Thornburg in Spotsylvania County. No forklift trucks existed at this time to pick up the entire stack, which is how it is done today.

    One vivid memory I have about the timber cart is that on Sunday afternoons when the mill was closed, my older brothers, W.O. and Fred, as well as their friends, would ride the cart down the track time and time again. It was a short ride, perhaps fifty feet, but it must have been fun. I never rode the cart as I was only four at the time and it was kind of dangerous. The reason it was dangerous was that the four-wheel cart had no brakes and just ran off the tracks on the downhill end if no blocks were put on the tracks first to stop it. They had to jump off of it before it got to the end of the track or they might be thrown off if they did not jump off first. After the ride, they had to put the cart back on its tracks and push it back to the uphill starting point for another ride.

What I was watching being done by my teenage brothers and their friends was a very short and crude version of the thrill rides that now exist at places like Kings Dominion and Busch Gardens. The coaster rides today have wheels that roll on a track just like the lumber cart, but do a thousand times more in dropping almost straight down, turning upside down and every which way. When the ride is over, people get back in line and ride it again and again, just like my brothers kept doing. Human nature never changes.

The sawmill was powered by a yellow engine which I'm guessing was a Minneapolis Moline gas engine. I would watch the sawmill running from on top of the little hill, which is where the shanties were set up. The shanty site was about two or three hundred yards from our house and was where the work crew slept during the week. They would arrive on Monday morning and live onsite until Friday afternoon-no commuting. One of the shanties was the cook's quarters, as the crew had one cook for the entire crew. The cook's name was Jack and he would get water from our well for their meals. My mother didn't like for me to go down to the sawmill, but I would go anyway, as even at that age (four) machines fascinated me. I knew not to stay too long or she would come looking for me and perhaps I would get the switch or at a minimum, fussed at. Jack would watch over me and sometimes give me small tidbits of food. The shanties were not very big, perhaps ten feet wide by twenty feet long. Other than the cook house, there were two sleeping quarters. The men ate outdoors and I'm assuming their toilet was in the woods. It was like camping out, something that people still enjoy today.

In my early years, trees were cut down either by axe or with a two-person crosscut saw. Chain saws weren't around then. Working a two-person crosscut saw in rhythm was an art which I never got the hang of. When one person was not doing it correctly, it was referred to as "riding the saw." I lost count of how many times I was told, "Don't ride the saw." After the tree was felled, the logs were cut in various lengths with a crosscut saw. These logs would then be "snaked" (i.e. pulled) by either a horse or mule to a roadway in the woods where they would be placed in a pile. This was done log by log. You would hitch an animal to the log with a harness consisting of a rather large padded collar that was placed around the animal's neck at its shoulders. Two wooded parts with metal straps attached to the wood (called "hames") were placed into grooves in the collar. The hames had metal rings attached to them in which two chains, one on each side of the animal, were run to a wood-and-metal bar behind the rear feet of the animal. These two chains, called "traces," were rather long, and the bar they attached to was called a "singletree." The singletree was wider than the animal's stance and had a ring in the center on the opposite side of the trace chains. From this center ring a short chain went to what was referred to as a "set of grabs." The grabs were two steel bars connected through a pivot pin. They had a ninety-degree bend on each end with a sharp point that attached to the log. The grabs were hand-set into each side of the log you wish to pull. When the animal moved forward, this pulled the short chain which in turn caused leverage action across the grabs' pivot point. This forced the points into the wood, which enabled the animal to pull the log. The animal was guided by the reins (usually two long leather straps) hooked to a bit in the animal's mouth. The bit was held loosely in place by a bridle strapped to the animals head. A person could walk to the side or behind the animal and by tugging on the reins cause the animal to turn or stop.

At the same time you were driving the animal with the reins, you were giving the animal voice commands. After two or three trips along the same path, the animal learned what he was supposed to do, and the driver could then wind up the reins and hang them some place on the harness. The animal was then guided solely by voice. "Gid-e-up" meant go forward, "Whoa" meant stop. "Gee" meant turn right and "Haw" meant turn left. Now think about it for a moment. Can you talk to your computer which may control a machine and cause that machine to start, stop, or turn one way or another? Perhaps somewhat with advances in technology, but nonetheless, it's amazing what God's creatures can do.

When making the log pile, a small log (typically fence-post-sized) was placed on the ground at right angles to the log pile, and the logs were rolled onto this cross log. This created a gap or space between the ground and the bottom of the log so that a chain or cable could be attached to one end of the pile of logs. The piling of logs operation in the woods could be a one-or-two-person operation. If it was performed by just one person, he had to walk with each log to the pile area and unhook the grabs. When two people worked together, one hooked the animal to the logs and the other remained at the pile area unhooking the grabs and rolling the logs into the pile. Of course, the animal had to obey both persons' commands. I wonder how many people today would know how to train an animal to do this log piling operation.

Modern logging is done by a person sitting in a heated or air-conditioned cab on huge machines. The modern tree cutting machine is large enough to approach a standing tree with a thick, spinning tooth blade head set just above the ground. I once observed one from a distance. The operator would hit the tree on one side, cutting off about half of it, then back up, turn the machine at an angle and hit the other side with the cutting blades. When he made the second cut, upper arms would grab the tree before it had time to fall. He would then back the machine up, carrying the entire severed tree over to his pile area. There he would lean the tree over the desired area on the pile and let gravity take over once he turned it loose. This entire process took less time than it would take for you to read about it.

Back to 1936, after the logs had been piled, they had to be transported to the sawmill to be cut into lumber. When you snake logs in the woods by dragging them along the ground, they got very dirty. This dirt would make the saw at the sawmill get very dull over time. It was best not to drag the logs over the ground any farther than absolutely necessary. These piles were moved to the sawmill by animal-powered log carts. These log carts had two huge wheels six to seven feet in diameter with steel tires that were about six inches wide around the rim of the wheel. The carts had an arched shaped axle so that they could be backed over the pile of logs which was about three or four feet high. A long tongue was attached to the arch axle which was used to hitch the animals to through their harness. A heavy chain was attached to one side of the arch which ran down to the ground. The other end of this chain was pulled by a rod under the logs. The gap that was created by the fence-post-sized pole enabled the chain to slide under logs and then be hooked back as tight as possible to the other side of the axle. When the animals pulled on the tongue, it caused the pile of logs to be picked up off the pole. As they moved forward, the whole load moved with the cart along the ground with most of the weight being distributed on the two giant wheels on the back of the cart. The team was then driven to the sawmill with the driver walking beside the front of the cart issuing commands to the animals and pulling on the reins. Once he arrived at the skid poles at the sawmill, the driver would back up the team to get slack in the chain to unhook the chain from the axle. The cart was then pulled forward, thus moving the chain from under the logs. The pile of logs was then rolled onto the skid poles, and then onto the carriage where they were cut one by one.

Modern day loggers use big four-wheel drive machines that have two large hydraulic grabber type arms. The driver backs the machine, called a "skidder," over the butt end of a pile of trees. He then uses the two large arms to squeeze the pile of logs together and then picks them up off the ground. Once he has the load firmly squeezed off the ground, he drives off dragging the remaining portions of the trees on the ground. The limbs are broken off if they hang on something. The trees are dragged to where the logs can be cut to length and loaded onto log trailers where they are hauled to large, semiautomatic computer-controlled sawmills.

Sawmill carriage in position with skid poles to receive logs on poles.

The main person at the sawmill was the sawyer. He determined how the log was placed on the sawmill carriage and had control over the functions of the mill. He has a helper called the log turner. The carriage was a long cart that had flanged wheels that rolled on a track like a short railroad. The logs were rolled up the skid poles and onto the deck of the carriage. The sawyer made the decision of how he wanted the log turned before he began to saw. He basically made a mental plan of how the log was going to be cut up before he started cutting.

A modern mill uses computers to scan the log and present the sawyer with two or three options on a screen. The sawyer then picks the one he likes best. The mill itself then does the remainder of the work depending upon how automated it is. With the manual mill, the sawyer and the log turner had to work together on the larger logs. They used long-handled stick devices with sharp hooks called "cant-hooks." With these, they hooked onto the log and could pull and maneuver logs of great size. The log was secured to the carriage, which was accomplished by bringing up moveable posts called "head blocks" against the outside of the log.

The head blocks had sharp pointed hook devices called "dogs" that were dropped down on the top of the log and were locked into place. The dogs kept the log from moving on the carriage. The front side of the carriage ran very close to the turning saw blade. The amount the edge of the log was pulled

Sawmill in operation. Note Sawyers left hand on Carriage direction lever.

beyond the edge of the carriage determined how thick a section is cut from the log. The sawyer did this by pulling a handle that moved all the head blocks in unison. The head blocks could be moved back and forth with this handle. In order to actually cut the log on the carriage, the carriage had to be slowly moved on a track to move the log through the saw. This was done by the use of wire rope cables, pulleys, and a wire rope drum that was powered off the saw mantel through flat belts that could be tensioned by the sawyer. The sawyer used a long handle that pivoted either forward or backwards. If he pulled back on the handle, he caused a flat belt to tighten which in turn caused the cable drum to rotate to move the log forward to cut. If he pushed forward on the handle, this caused another flat belt to tighten, rotating the cable drum in the opposite direction and thus backing up the carriage. When the handle was in the center position, the carriage was in neutral. The person that caught the timber from the log on the opposite side of the saw was called the "off-bearer." The off-bearer had to keep the timber or slabs out of the way of the saw. It usually took two or three off-bearers to keep up with the wood that was cut, to sort it, and pack it for loading.

THE SAWMILL, LOGGING AND THE FORESTS

The 1936 sawmill did not have a sawdust carrier chain. In the 1930's, labor was cheap and there was very high levels of unemployment. Thus, lots of jobs were done by hand. Back then, we paid our farm laborers $1.00 a day, just to give you an idea as to what it was like. They also got a noon meal, we called dinner. One person was required to keep the sawdust out from under the saw. To do this, he used a six-foot long handle scoop shovel and a wheelbarrow. He would load the wheelbarrow with the shovel; the long handle enabled him to stay back from the turning saw. After he loaded the wheelbarrow, he would put down the slabs with the flat side up to make a narrow walkway over the soft sawdust. The sawdust was piled approximately three feet deep and eventually covered an area as big as a house. It took a crew of twelve-plus men about two months to complete the timber-cutting. After the sawmill was gone and all was quiet, I would go down to where the sawmill had been and play on the sawdust and slab piles. Thus ends the Allen sawmill memories!

One of the regrets I have about the forests we have now is that the entire countryside has been cut in Virginia. It would be nice if, say, 1,000 acres or 100 acres or even 10 acres had been marked off and set aside 400 years ago and no one allowed to cut or take out any tree or wood whatsoever from the area. If this had been done, then we, and future generations, could visit the site and get an idea of what virgin forests were like in 1607 when the Indians had the entire country, prior to the arrival of white men. The site would have to have walking paths so people could walk among the trees. Such an area can be found in the Smoky Mountain National Park in eastern Tennessee near Gatlinburg. There, three and one-half miles one way up a hill on the side of the mountain lies Allbright Grove. Allbright Grove has never seen a saw, an axe, or any wood removed from the land. In the fall of 2007, my wife Shirley and I took a day hike to this grove. Some of the larger poplar trees are over six feet across at the bottom and over one hundred feet tall. The forest is thick like a jungle with a mixture of small trees, plants and bushes as well as some huge trees and everything in between (including dead and broken trees). It's as close to an example as you will find of what a forest is like when man does not do anything to it.

The home of James Madison in Orange, Virginia, has a woodland area that is very close to virgin. While visiting the area, I learned that it has been what is called "selectively cut" as you can see a few stumps where trees have been cut. For the most part, this James Madison forest is like the original with only a few trees having been removed. Like Allbright Grove, it has some extremely large trees as well as dead and broken ones, which contributes to its jungle-like appearance. The Harris farm Cherry Grove had a small area like this on its west side between the back field and the North Anna River. Route 522 was cut through this strip of woods.

In the early 1940's, I remember walking with my father through this stretch of woods. This strip of woods had a few of what we call "old growth pines." These were pine trees three to four feet across at their base and around one hundred feet tall with no limbs on the bottom half. Their tops or crowns were very small. They were tall, skinny, and mostly straight trees which I believe are called "short leaf pines." Unlike the Virginia pine, these did not have many limbs. At the time of our walk, one of these huge pine trees had blown over and was lying flat on the ground. The log near the base was taller than I was at the time, as I could not see over the log. I climbed up on the stump by catching hold of the broken-off roots and got on top of the huge log. As I started walking down the log towards what used to be the top of the tree, I was stung by some sort of bee or wasp on the ear. I instinctively slapped my hand to my ear, killing the bee, but I never saw what stung me as its remains fell to the ground below. The pain was something terrible. I jumped off the log and took out for home, crying all the way from the terrible pain, leaving my daddy behind. I remember the pain to this day. I don't think anyone could have salvaged that tree, as the logging machines and saws of that time would not have been able to handle a tree of that size. I expect that tree was nearly four hundred years old. None of these pines are left there now. Don't the scientists claim that trees take pollution out of the air and give off oxygen? Yet we treat the forest and trees like trash when they sustain life. Shame on us for doing this.

CHAPTER NINETEEN

NEW ROADS AND BRIDGES

The road that served our farm was Route 612, now called Monrovia Road. This was before they built the new section of Route 522 through the back of the farm in 1949. After they finished the new section of 522, we still used Route 612. Route 612 was a dirt road with gravel hauled in from time to time to fix holes or bad sections as they developed. Gravel was a mixture of sand and small rocks dug up from the sides of river banks or out of the stream bed. All you had to do is load it on a truck. Most of the time, it was hauled from the Fredericksburg area where it was deposited by the Rapphannock River. Most of the rocks put on roads today come from a rock quarry where the solid rock is blasted out of the earth, then ground up to various sizes and put on top of the dirt to make a road. I can remember one winter when three sections of 612 got so bad that people got stuck in the ruts that were made from the traffic over the wet dirt. For two weeks the school bus could not come up from what is now Route 719 and I had a two-mile walk one way to catch the school bus. The reason the ruts got formed was when the dirt freezes solid in the winter, then thaws out when the weather warms up, it loses strength and becomes very soft and muddy. Soft dirt cannot support much load, and the wheels of the vehicles press down into the dirt and form a rut or track. Once you get in one, you don't have to steer the vehicle as the rut guides the vehicle like the rails do a train. When the ruts get so deep that the bottom of the vehicle drags the ground, you get stuck.

Route 612 was classified as a secondary road back then and still is. In those days all secondary roads were dirt or gravel roads. When it rained, they could be slick and you kicked up mud which got all over the vehicle. When they were dry, you kicked up great clouds of dust, which again got all over the vehicle. Driving over a dirt road takes much more skill in driving than driving on a hard surface road. Depending on conditions, you can slide much more easily on dirt. It takes more skill to drive on the varying conditions of a dirt road, versus a hard surface road.

At that time the only hard surface roads were the primary roads which ran between the cities and towns. Some of these roads were poured concrete roads. A few were asphalt surface roads, but most were what we called a "tar" road. A tar road is made by spraying hot tar over the dirt road; then while the tar is still hot and soft, a layer of crushed stone or pea-sized gravel is spread on top of the hot tar. These rocks or stones are then run over with a wide roller that presses the rocks into the soft tar. When the tar cools within one to two hours, you can drive over the road with no dust or mud problems. This lasts for a few years, but with time the tar developed cracks and then begins to break up. Then the process has to be done over again.

One main thing the tar or any hard surface does, which most people don't realize, is that it keeps the road dry underneath. Dry dirt can support much more load than wet dirt. To make a really strong road, one must cover the entire roadbed with fairly large rocks about eight to twelve inches deep, then put a hard surface on top of all these rocks so that no water can get in to make the dirt soft under the rocks. Much of the cost of building a road is putting in this hard material over the dirt roadbed and sealing it up.

In mid-summer of 1947, the Virginia Department of Highways (always referred to as "the state") began construction of a new two-lane highway through our community. This new road began at Ranny Jones' store near the border of Orange County. The road ran through the back field of our farm in Spotsylvania County, then crossed over into Louisa County. There the road joined with the existing highway, Route 522, near Ed Syrkes' store (where Route 719 begins today). The new highway was a total of six miles. The purpose of making this new section of Route 522 was to eliminate three one-lane bridges as well as some very hilly and curvy sections, and to shorten the distance between its beginning and end. The road still required three bridges; however, these new bridges were two lanes instead of one. The problem with the one-lane bridges was that if you came to the bridge and someone was crossing it from the other direction, you had to stop and wait until they passed before you could continue on. One-lane bridges are inherently dangerous and required that you to approach the bridge slowly as you never knew if someone would be on the bridge. Needless to say, they made your trip slower.

One especially bad bridge that was bypassed by the new road was known as Dillard's Bridge. Dillard's Bridge crossed the North Pamunkey fork of the North Anna River. It was the longest of the three one-lane bridges and was unusual in that it was built on a curve. In addition, the approach to the bridge contained very sharp curves on each end. These two factors resulted in numerous accidents,

Original "Dillards Bridge" on left. It had a wood floor and a very weak side rail system. The new bridge under construction was replaced by the present bridge over the exact spot as the original one when Lake Anna was formed in 1972.

especially for those individuals that were unfamiliar with the road. Our school bus had to cross this bridge both to and from school. As we crossed the bridge one afternoon, there were about six or eight sailors in their white uniforms standing in a line where the guardrail was supposed to be on the south bank. The approach on the south side of the bridge had a high bank on one side with a guardrail from the bank where the land dropped steeply down. As the bus slowly passed by the line of sailors, we looked out the windows and noticed that their vehicle had run off the road. All that was visible of the vehicle was its back end. Thankfully, none of the sailors were injured.

I remember my oldest brother, W.O., telling a story of being stopped by a tractor trailer driver and being asked how many more bridges he had to cross like the one he had just been over. I think Dillard's Bridge scared him. This new section of road greatly upgraded a hazardous section of Route 522. I'm confident that many lives were saved as a result of this new section of road, as motorists could now travel with more sense of security (as well as more speed).

Just about every day, you can read in the newspaper or hear on the news of someone, somewhere, being killed on the highway. I'm confident some of these deaths could be prevented by building wider and straighter roads than the ones in use today. Our leaders are failing us in not recognizing the need to improve our highways by widening and straightening our roads. Sure this costs money, as everything does, but what is a person's life worth? Don't make the mistake of saying or thinking "It won't happen to me!"

The survey work for the road began before 1947, lasted about two years, and resulted in several different routes staked out. Groundbreaking began in midsummer 1947 on Mr. Tesdale Burruss' farm. Any woods that the route was designed to run through were cut down by hand. The many culverts or pipes that were needed to allow water to run from one side of the road to the other were all laid by hand. All of this backbreaking work was performed by convicts using hand tools such as axes, saws, picks, and shovels. A convict crew consisted of about a dozen men, along with one armed guard. The convicts were brought to the work site in what resembled a box set in the back of a dump truck. While at the work site, the convict box was removed from the back of the dump truck so that the truck could be used to haul materials throughout the day. At the end of the day, the box was put back into the back of the dump truck and the convicts were taken back to camp. The armed guard would typically stay about eighty to hundred feet away from the convict work crew. His sole purpose was to observe them work and to make sure they did not attempt to escape.

A state employee directed the convicts' work. In addition to road construction, the convict work crew also built the required fences along the side of the road. The new road required an eighty-foot-wide right-of-way strip of land. The road itself was thirty-six feet wide from the ditch center on either side of the road. The paved surface was twenty-four feet wide. The original machines they began construction with were an International TD14 Bulldozer, an Allis-Chambers HD-14 "crawler," a motor grader and a Caterpillar "TEN" (a very small track tractor with a gas engine). I remember thinking that the TEN was not a very impressive piece of equipment at the time, but today it is a very valuable antique. The TEN was used to pull a "sheep's foot roller." This roller was used to hard- pack the dirt in the fill area. The main tractor, the Allis-Chambers HD-14, was attached to a cable-controlled-earth-mover called a "pan." A Pan was used to scrape up the dirt as well as to move this dirt to the "fill" where it was spread out in a layer which was then packed by the roller.

The motor grader was used for the precise shaping of the road. A motor grader had a long curved blade near its center between the front and rear wheels. The blade could be positioned in various ways in order to make smooth, gradual banks and curves. After several months, a new pan and a dark green Caterpillar D7, with a cable blade, were brought in to replace the TD14. The D7 was a World War II army tractor; thus, it was not the Caterpillar traditional yellow. The D7 was a slow machine, but was very powerful with a weight of around twenty tons. The D7's engine's top speed was around one thousand rpm. The engine would pull down low but never stall while under load. I observed it being used to dig out stumps. Sometimes the operator would load the engine so much that you could count the exhaust "pops." When he would disengage the clutch, the engine would rev back to speed after about thirty seconds. By comparison, modern diesel engines idle at around one thousand rpm and turn up twenty-two hundred to twenty-five hundred in top speed but can be stalled too quickly. The HD-14 could literally run circles around the D7, but it had its own set of flaws. Around every two months or so, the HD-14 would be parked needing repairs. Perhaps a bearing went, a gear broke, or the steering clutches required maintenance. The HD-14 traded reliability for speed when compared to the D7. The D7 was a slow, steady powerful workhorse that I never saw requiring maintenance while on site. It was a model of dependability. In my opinion, the Caterpillar machines earned a reputation for being dependable, so people knew they could be trusted. It's very frustrating to have perfect conditions for construction, but have to wait on broken or unreliable machinery. Most people will not stand for unreliable equipment and thus acquire dependable machinery. In my opinion, Caterpillar's reputation for reliability is a dominating factor as to why the company still exists today as opposed to International and Allis Chambers.

 The technology at this time used over-center manual operated clutches in machines of this kind. They all had a clutch lever on your left hand side when you sat in the seat on the machine. You pulled back on this lever to engage the clutch. This clutch locked up when you pulled it over center, so that you did not have to constantly hold it to make the machine move. All the crawlers at this time were like this. You pushed the lever forward to disengage the clutch and pushed the brakes with your feet if necessary to stop. This clutch has been replaced with torque converters. Now you have a small lever that you move with your left hand that you use to select the forward or reverse direction plus a park position. You also select the speed or gear you want with the same lever very similar to the select lever on an automobile. It took a lot of pulling and pushing with your left arm on one of these clutch tractors. This would get you tired before the end of the day.

Crawler tractors have no air cushion between the machine and the ground, such as you get with a tire. If a crawler is run over rocks, hard or frozen ground, or pavement, you feel a lot of vibration and roughness. When it is run over soft dirt, the ride is not bad. It is my belief that if one spends all his working hours for years and years being constantly shaken by the machine, it will cause some kind of health problem to develop. God did not design a person to be constantly shaken, just as he did not design us to breathe tobacco smoke for years and years that we know for certain will cause some kind of health problems to develop.

When the crawler tractors were first made, all they could do was pull something by their drawbar. Mr. R. G. LeToureau is credited with mounting a heavy steel blade in front of a crawler tractor. He raised or lowered the blade with a wire rope cable powered by a winch. The cable was hooked to the top of the blade that was in front of the tractor. This cable was run through a pulley that was mounted high up above the top of the tractor, and then to a winch that was mounted on the back of the machine. The winch got its power from the tractor engine through a hand-operated clutch. A crawler tractor equipped like this is called a "bulldozer." It gets its name because just like a bull (the animal) who uses his head to push something, the crawler tractor with a blade in front can push something. Mr. LeToureau was a very creative person who invented many new machines for moving dirt, rocks, and other heavy objects. He never did accept hydraulics to raise or lower his devices. He claimed they always leaked oil, which they did at that time. This hydraulic oil-leaking problem has largely been solved now, with much improved seals and tolerances. Very few machines now use his cable method to raise or lower something. In his autobiography,

R.G. LeTourneau: Mover of Men and Mountains, by Moody publishers, he tells of how he got his ideas for the many earth moving machines he invented. He was the first to put wheels on a pan. He was first to put the motor on the pan and make it into a self propelled machine. The first to use air filled rubber tires instead of steel wheels on earthmovers to name a few of his inventions. Get the book and read of all the struggles and heartbreaks he had and of his Christian beliefs and being a partner with God. The next time you get on an interstate road, you can thank Mr. LeTourneau and God for giving him the ability to think of and then actually build machines that can move the earth to make the interstate road possible.

The most difficult skill to learn to operate a bulldozer is how much to raise or lower the blade. When you are sitting ten to fifteen feet behind the blade, you can't see just exactly what it is doing. There are several different things you can observe. You can see the bottom of the backside of the

Allis-Chalmers bulldozer in operation. 1962

blade and maybe see what is rolling up in front of the blade if enough material is piled up to come above its top. You learn to listen to the sound of the engine as well as feel the strain being put on the machine by the load in front of you. With time, one can learn how to work the blade as if it were a shovel in your hands. It's not as easy as it looks.

One thing that I am in awe of is that back in the 1800's when they built the first canals, and then the railroads, it was done by hand and animal power. They had to use the hand axe and saw to cut trees out of the way. All the digging and earth moving was done with picks, hoes, shovels, and carts and wagons. Imagine digging a ditch forty to fifty feet wide by four to five feet deep for miles and miles by hand for the canals. Incredible. With the railroads it required cuts and fills to make what they call a "grade," which is a continuous slope or plane. These were all done by hand then, because crawler tractors did not exist. With the use of these powerful machines, one person can do in one day what would require 1,000 people with hand tools months to do. Have we made good use of this incredible leap in productivity with these machines?

One very obvious thing about these crawler tractors then and now is the operator's seat. Look at a picture of a tractor in the 1940's to 1950's, and the operator is sitting on top of the machine with nothing but the sky overhead. No protection whatsoever. Sometimes you would see a large umbrella mounted to keep the sun off the driver.

D8 Caterpillar pulling a sheeps foot roller in Richmond, Virginia, building what is now I-95. Note absence of operator protection. 1957

Now they are required to have rollover protection (ROPS). This is a heavy massive structure mounted to the tractor that is strong enough to support the tractor in the event of a rollover. It has to be built strong so that it will protect the operator whose seat is within this structure. Many people have been killed on open seat crawlers, mostly because something fell on them. This was especially true when the tractor was used to push over trees. If you have to have this structure protection, then you might as well enclose it with glass panels and put in heat and air conditioning and give the operator some comfort. This is what the modern tractors have come to.

NEW ROADS AND BRIDGES

The construction of the bridge that crossed the same stream as Dillard's Bridge was handled by the Abernathy Construction Company from Glen Allen, Virginia. They built a three-section, two-lane wide poured-in-place concrete bridge. (This bridge was later replaced by a longer, higher bridge around 1970 with the construction of Lake Anna). The primary bridge that crossed the North Anna River was on Uncle John's property on one end and on the Estes property in Louisa County on the other. This bridge was built by the state and consisted of concrete poured columns, steel trusses, and a wooden floor. This bridge has since been replaced as well by a higher, longer bridge with the construction of Lake Anna. The last bridge ran across Christopher Run on the Bazzanella property. This bridge was a poured-in-place concrete bridge as well and was constructed by the state. It too has since been replaced with the construction of Lake Anna. Once the bridges were finished, the new road was opened for traffic in 1951.

While on the topic of bridges, another bad bridge I remember was on Route 522 over Contrary Creek. This was a one-lane bridge of about thirty feet long. What made the bridge dangerous was that the approach road coming from Mineral when traveling north was curved so that when you exited the curve you were immediately at the bridge. Since it was one-lane, if someone was on the bridge, you had to stop to let them pass. The curvature of the road made visibility a problem; thus the bridge was dangerous particularly for those that were unfamiliar with the road. Around 1940, a two-lane bridge was constructed adjacent to the existing bridge to replace it. This second bridge was safer because it was higher, wider and had two-lanes. The curve that preceded it when approaching from the south was still an issue. As a result, a third bridge was built downstream of the second bridge. This bridge was again higher and wider which contributed to greater visibility, but again many accidents occurred on that section of road. Finally in the 1990's, the entire section of road of Route 522 from the town of Mineral to Dickinson's Store was rebuilt. A fourth bridge was constructed that was yet again higher, wider, and longer. This fourth bridge was constructed downstream of the prior three bridges. Finally, they addressed the hilly, curvy road itself that contributed to the majority of the accidents. You can easily drive 70 mph through this section of road (but be careful not to get caught speeding).

It would be nice if all of our highway and road systems were built like this new section of Route 522. The problem is that our leaders don't seem to understand (or care) that our road system was designed over sixty to seventy years ago and is inherently dangerous for modern vehicles traveling at greater speeds. People get killed on these outdated roads and it is my opinion that their blood is on the hands of those in power that could but would not do something about the situation. Who knows, one day they might lose a loved one due to an outdated road which will inspire them to take action to improve the safety of our often outdated road system.

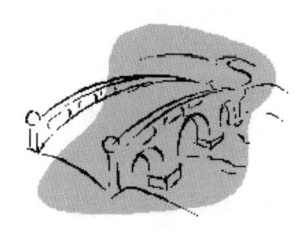

CHAPTER TWENTY

Pedro

As a farm family, we were members of the Farm Bureau, an organization for the purpose of promoting the interest of farmers. It lobbies to influence national, state and local governments to enact policies and laws that benefit the farming community. In addition, it offers a variety of insurance services. In 1955, the Farm Bureau had a national program that gave young people from foreign countries the opportunity to live for ten months with an American farm family. The object of the program was for the cross-education of young farm individuals in other countries. By living with an American farm family, they would learn not only our farming methods but also gain first-hand experience of our culture and how we lived our lives in general. I believe it was an exchange program whereby young people from the US could go to other countries and learn from them as well.

My mother, my two brothers, and I (operating then as the Harris Brothers) applied to the Virginia Farm Bureau Federation to participate in the program. My father had passed away in 1953 and my middle brother, Fred, and I lived at home with our mother. W.O. had his own home and family, so we had an extra room in the house for a long-term guest. We were accepted to participate in the program and in mid January of 1956, Fred drove to Richmond to pick up Pedro Augusto G. Bostos of Minas Girias, Brazil, South America. The name "Minas Girias," Pedro told me, translates to "general mines." It's ironic how he came from an area in South America that had mines of various kinds of ores to a similar area in North America called Mineral that was known for mines of various ores as well. Within a six-mile-radius of the Town of Mineral, I have counted seven different mine shafts which produced sulfur, zinc, lead, iron, and gold. One of the shafts was named "Arminos" which means "all metals." Mineral was originally named Tollersville, but its name was changed to Mineral after the discovery of the many ores in its general vicinity. The ores are still present there today, but the mines have been closed, as other locations were discovered that produced either higher quality metals or were cheaper to mine.

When Pedro arrived, he had completed college where he had studied English. His native language was Portuguese, of which we had no knowledge. At first it was difficult for both him and us to communicate. It was like he was in deep water and had to learn really quick how to swim, otherwise he wouldn't make it. He knew enough English to start with, but the first two months were very difficult for him. We learned how to converse with him, because when you are in nearly constant contact with another, you learn to communicate differently than you would with those you are familiar with. Thankfully, Pedro was highly intelligent, a fast learner, and a natural leader.

In addition to the above qualities, Pedro was an excellent swimmer. I remember when a group of us went to Overhill Lake near Richmond, Virginia, Pedro realized that I was struggling and got me into shallow water. I never learned how to swim. I possibly could have drowned in that situation without his assistance. Another of Pedro's traits was that he charmed all of the women from ages eight to eighty. He looked like Clark Gable, and girls couldn't help but idolize or fall all over him. Like us and the hired help, Pedro worked on the farm, although I do not remember his wage. He operated some of the machines, helped with the milking, and generally observed all that we did. Once he got my mother to purchase some whole coffee beans, and he made what he called "genuine coffee" by going through the same process as he did back home. It was so strong we couldn't drink it; it didn't even taste like the coffee we were accustomed to. This was one of the things we learned from Pedro about his country and culture.

Pedro was always pointing out the potential for a glorious future for Brazil. He spent time listening to Brazilian music, he wrote to his family, and of course, he received mail from them. While here, he collected numerous mementos of his trip to carry back to Brazil. On his return trip by ship out of New York, he later wrote me that most, of his luggage was stolen. He was deeply disappointed about this, and who wouldn't be? While here, he acquired a 35-millimeter camera and snapped many color slides of the farm and things in general. He took photos of the Belmont Home Demonstration Club, of which my mother was a member. He put on a program for them about his country as well as for the local Ruritan Club. I remember him taking a group photo outside of the Kirk-O-Cliff Presbyterian Church that we attended. Pedro was Catholic, and once I took him to a Catholic Church in nearby Orange County. I had never attended a Catholic service; thus I felt like a fish out of water as I didn't know when to kneel, sit, or stand. Some of the service was in Latin, which I didn't understand. As a result, the whole experience didn't do much for me, but I hoped it meant something to him.

His many color slides made me realize the beauty of my surroundings, and after his departure, I purchased a camera and began taking slides. Pedro greatly enriched my life and I will always be grateful for what he taught me. We have kept in contact by mail since his ten-month stay in 1956. Upon his return to Brazil, Pedro eventually became the manager of a milk plant in Rio de Janeiro. In 1977, I had the opportunity to visit him in Rio. I met his family and toured the milk facility he managed. Pedro claimed that his brief stay with us influenced him into selecting his career in the dairy field.

What the world needs is more of this exchange of young people between countries and cultures to learn first-hand how each other lives. How many of us understand what life is like in an Arab country? How many, Middle Easterners and others understand how we live? An exchange program that promotes living ten months out of a year with another family is an excellent way to promote peace between countries and cultures.

CHAPTER TWENTY-ONE
Grandpa Abner

The name "Harris" is very common. To trace my descendants back to England, we have record of at least six people with the Harris name invested in the Virginia Company. The Virginia Company had over one thousand shareholders, of which very few actually came to Virginia. The motivation of this company was to establish a colony in Virginia for economic gain. The record of the ship *Prosperous* lists the name of one Thomas Harris, age thirty-eight, who came to Virginia in 1611. The records also state that one Adria Osborne, age twenty-three, came to Virginia on the ship *Marmaduke* on November 23, 1621. The two became a couple and married by the year 1623 and patented lands called Neck-O-Land in Henrico (presently known as Curles Neck). The early settlers would just go out and mark off an area of woods by placing chop marks on trees and then claim it by a piece of paper called a patent. At one time this was the largest farm in Virginia. This Curles Neck property is very valuable today.

Adria Harris died soon after a son, Robert Harris, was born. Robert was taken back to England where he grew up. Thomas didn't waste any time in finding another wife. By 1625, Thomas had married Joan Gurgany. and a daughter, Mary Harris, was born in 1625. The couple had two other children, both boys, Thomas Harris, Jr. and William Harris (born in 1629). The presentday Harrises can all trace back their lineage to Thomas Harris. Information on Thomas Harris can be found in the W. T. Baker, Sr. book entitled *The Baker Family of England and of Central Virginia*. In the book beginning on page 242, Baker has a chapter on the Harris family. Baker features genealogies of many early Virginian families along with the Baker family.

GRANDPA ABNER

Now, move forward eight generations to my grandfather, Abner Nelson Harris, Jr., who was born on April 17, 1835. Grandpa Abner was married twice. He married his first wife, Mary Catherine Kimbrough, in 1857, she later died during childbirth, as well as the child. Grandpa Abner, like Grandpa Dick, was a soldier in the Confederate States of America (CSA) army. Grandpa Abner had two brothers, John and Frederick Harris, who were also soldiers in the CSA army. John and Frederick were both killed in separate battles in 1862. Grandpa Abner was badly wounded in his stomach during some battle during the Civil War. It is thought that he was able to survive this wound to the stomach because he was near starvation or on a starvation diet.

Grandpa Abner had a reputation of being able to nurse people back to health. He nursed a fellow soldier under very primitive outdoor camping conditions, saving his life. In any event, he survived the war and was left with essentially nothing besides his life, which is the most valuable asset anyone has. Nothing on this earth can compare in value to it. This applies to every living person and don't you ever forget it.

In 1866, Grandpa Abner married his second wife, his cousin Victoria Martha Harris. The couple moved west, and he spent a few years after the war working as a farm laborer in Hopkinsville, Kentucky. He was looking for a better life in what was then the West. Grandpa Abner moved back to Virginia and purchased the property called River Bend, near Lahore in Orange County, Virginia. Grandpa Abner and Victoria lived at River Bend from 1872 to 1880 where my father, William Overton Harris, was born in April 1873. The couple had a total of nine children, eight of whom (including my father) lived past childhood.

In January 1881, Grandpa Abner moved his family to Clifton in Louisa County, Virginia, where he rented a farm. I always knew this as the Bazzanella place as Mr. Max Bazzanella, who emigrated from Austria in 1913, purchased this farm. The Bazzanella family still owns this property.

In 1884, Grandpa Abner moved across the river to Cherry Grove. They rented this property and farmed it until June 1892, when they moved to Glenora renting and farming it until 1899. Glenora is now part of the state park located on Lake Anna. Their last child, Aunt Lucy, was born in 1887. It is very difficult for us to imagine what a struggle life was for people of the South after the Civil War ended in 1865. It took nearly 100 years for the South to catch up with the rest of the United States.

Grandpa Abner died in November 1899. Between then and July 1900, his widow, Victoria Martha Harris, moved back to Cherry Grove and agreed to purchase the property. By this time, the oldest son, Abner ("Unk"), was thirty-two years old, and the youngest son, Uncle John, was fifteen. My father was twenty-seven years old, and all big enough to do farm work with the knowledge of how to work and manage a farm. I do not know the terms of the deal, but I do know, like with most property purchases, they didn't have enough money to pay for it in one sum. Before 1915, my father, Uncle Frank, and Uncle Abner ("Unk"), had taken deeds to portions of the property by paying off what was owed on what they took ownership of. My father and Uncle Frank took approximately 100 acres each. Unk took approximately 150 acres. Uncle Frank's land was the northernmost part of the 600 acres, and it ran to the west of the road from the Kirk Church to the Cherry Grove house. Daddy's part was to the west of this road and in the middle of the tract. Unk's part was from the river up to Daddy's land and to the west of the house. The North Anna river was the southern border of the farm. Grandma Victoria's will left the rest of the property, including Cherry Grove to her single children who never married. She or they would have to have paid off the remaining debt that was owed on the property. The single children included Aunt Lucy, Aunt Virginia ("Virgie"), Uncle Abner ("Unk") and Uncle John. I am not sure how Aunt Mary Caroline ("Tannie") and Aunt Victoria ("Too") got their shares of the estate. They used nicknames a lot back then, and sometimes it took years before you discovered what a person's real name was. My father's nickname was "Willie."

Daddy rented Uncle Frank's land and farmed it. Uncle Frank was the C&O Depot agent in Mineral, Virginia, and lived in that town. Daddy eventually bought Uncle Frank's land which was formerly owned by my deceased brother, Fred. Daddy built his house in one corner of his property in 1915. Brother Fred lived in the house up until his death on January 12, 2009. The dairy complex is now run by his only son, Charles. My oldest brother, William Overton ("W.O.," who died December 29, 2002) was willed all of the land on the west side of Route 522, a road that was cut through the farm in 1949.

W. O. Harris Family 1934
Back Row - (l - r) Louise, Lassie, Mother, Dick, Father, Vicky
Center Row - Fred, W. O. Harris, Jr.
Front Row - Virginia Todd and 2 dolls (Uncle Henry's daughter)

CHAPTER TWENTY-TWO

GRANDPA DICK

My earliest memories go back to the summer of 1936. I was born on March 3, 1932, in the house my father had built in 1915, located in the southwest corner of Spotsylvania County, Virginia. I am told that a couple of days after my birth, they had a terrible snow storm or blizzard that marked the occasion. My mother named me after her father, Samuel Dick Todd. Grandpa Dick was born on September 26, 1846, and died on February 24, 1935. While still a teenager, at the age seventeen, he enlisted in Company E, 1st Regiment, Virginia Reserves Confederate States of America (CSA). I have no memory of him, but I do have pictures of me sitting on his lap.

One of the most remarkable things Grandpa Dick did, in my opinion, occurred when he was twenty-six years old. In the fall of 1872, he set out on a journey, riding on a horse, from Caroline County, Virginia, to Texas; he returned after one year's stay. The reason for this journey was to claim his inheritance of property that his father, Samuel, had left to him and his mother in his will of 1866. His father, Samuel Todd, left him and his mother property in Jasper, Robertson, Rusk, Fort Bend, and Warton counties in Texas. The property in the five counties amounted to nearly ten thousand acres. Some of this property was left to Samuel Todd's first set of children.

Samuel Todd was married twice. In his first marriage, he had five children (three girls and two boys). He had a son named John, who was taken prisoner by the Union Forces during the Civil War. John died in prison in either 1864 or 1865. Samuel Todd wrote his will in 1866, a few months before he died. His first wife's name is almost a mystery, as there is no record in the Melville M. Todd 1979 book as to her birth, marriage, or death. They do list the names of their five children. They are unsure if her name is Sarah or Susan Holman. Samuel Todd evidently did not keep in contact with his children from the first marriage very well, but he mentions a son, James H. Todd, in the will and states he does not know if he is living or dead.

His second marriage was to Ann G. O. Dick in 1845. They had one child born in 1846, my grandfather, Samuel Dick Todd. Thus my name is "Dick" and not Richard. The couple resided in Fayetteville, Tennessee, but Grandpa was born in Clinton, Mississippi. Samuel Todd owned property in Louisiana, Mississippi, and Texas in 1845. In researching the Todd family history compiled by Melville M. Todd in 1979, I discovered that Ann and Samuel separated when Grandpa was about two years old. Ann and two-year-old Grandpa moved back to her original home place, Twickenham, in Caroline County, Virginia, where she lived for the remainder of her life. Grandpa grew up at Twickenham. Samuel stayed in Tennessee, but eventually moved to Virginia near his wife, but not with her. In a letter to Ann at Twickenham, Samuel tells about not being strong enough to travel, and his doctor advised him not to travel to Virginia at that time. He must have had some kind of illness and was on a slow recovery. As mentioned, Samuel eventually did move to Virginia, as his son, John had addressed a letter to him at Chancellorsville, Virginia, which is not very far from Caroline County, Virginia (twenty to thirty miles). His will is recorded in Hanover County, which joins Caroline County. Samuel Todd passed away in 1866.

Grandpa's horseback trip to Texas was not successful in his getting a deed to any property in Texas. We have letters written to him by his mother, Ann, that he saved and brought back to Twickenham. In this correspondence she responded to questions he must have asked in his letters to her concerning the property. She had not saved his letters like he did hers. We have letters from his half-sister, Elzira Greer, of Petersburg, Florida, and from his niece, Juliette Williams, of Ocala, Florida, that he saved. These letters were written in 1886 and 1887, and they contain discussions about the Texas land. Thus thirteen or fourteen years after his horseback trip to Texas, he and his family were still trying to get the title to their property in Texas.

It's my belief that Grandpa Dick was unsuccessful in acquiring his inheritance due to a number of factors. Most likely the land was never properly surveyed; it was not properly recorded in a courthouse (if the counties had a courthouse built then) and the fact that Samuel did not live on any of the properties. The Homestead Act did not exist in the 1840's. From 1836 to the end of 1845, Texas was like an independent country. It joined the United States on December 29, 1845, as the State of Texas.

This would have been an absolutely wild and unsettled country when Samuel went there and made land deals sometime before 1845. Protection of one's property then was done by the gun and, since he wasn't there, he didn't do that, either. It turned out to be a lost cause for Grandpa, but we cannot help wondering what could have been. If Grandpa had been successful in claiming the land, he probably would have moved there. Yes, perhaps some of that land could have had oil under it, but our mothers and fathers would have been different and the present generations would not exist as we know them today. Did God plan for it to work out the way it did?

To me, the decisions we make and the things we do or don't do affect history. We never think about how the things we do today may affect the unborn 100 years from now, but they do. Do we make our decisions based on God's word and laws? Do we study God's word and laws and worship him by becoming affiliated with an organized church or religion? So many people don't even believe in the existence of God or a higher power of a spiritual nature. In fact, we have trouble in understanding how a spirit can exist. Where do you get your guidance as to what is the right way to live your life and what is the wrong way to live your life? Only by making a conscious effort to learn what is right and wrong can we know what is a right decision; it's not going to fall in your lap. If we try to do what is right, then we can be confident that things will turn out as they were supposed to by God's plan.

Like his father, Grandpa's first wife died leaving him with five small children. Ten years later he married again and had two more children. One child was my mother, and the other was her brother, Lindsey Marshall Todd. It's ironic how it turned out that both father and son both lost their first wife and they both remarried and had more children. When Grandpa's first wife, Sallie, died at the age of twenty-eight, he was left with four small children ages two to seven, and a two-month-old baby. Sallie's mother, the children's grandmother, Mary Frances M. Hansborough, took four of the children and raised them. Her sister, Evilina M. Dickinson, took one and raised her in her home. Did baby formula exist then? How would a grandmother nurse the two-month-old baby? This baby was our Uncle Henry. This had to be a very sad and stressful time for the whole family. However, they lived through it and became productive citizens. I knew and remember each one of them.

Grandpa Dick was a religious person. He worshipped and worked at Trinity Baptist Church, holding the position of treasurer, when he lived at Benvenue in Louisa County. When he lived at Ridgeway in Spotsylvania County, he went to Mt. Hermon, the nearest Baptist Church. In his last years, he moved to the town of Louisa and attended Louisa Baptist Church. Some people suggest I inherited my talent to make things from him, as he had the reputation of being a skilled carpenter. I am thankful for the good genes that have been passed on to me.

CHAPTER TWENTY-THREE

A TESTIMONY

I would like to share with you the story M. B. Ott told to me. It is about his out of body experience that occurred around 1930. Some background about Mr. Ott. He was one of my first customers when I started to build machines to feed livestock. He and his sons made a visit around 1965 or 1966 to see a new type bunk feeder that I had begun to build. It was different from all others in that it used a moving wide belt to feed cows in a barnyard. A bunk feeder replaced the manger that was built into the dairy barn, as described in the dairy chapter of this book. Bunk feeders were a long feed trough set up in the barnyard lot outside of the barn. They used mechanical means to distribute the feed to the animals. The animals were able to eat from either side of this trough. The feed was stored in a tower silo and was removed each day into the feed bunk. Electric power was used to distribute it the length of the trough. It was a step in the evolution of feeding cows. Mr. Ott lived on a farm several miles from the town of Remington, Virginia. I made a sale to him, and over the years he and the sons bought four of these feeders from me.

As Mr. Ott aged he got to become the "Go For" person for his sons, as he would come to Mineral several times a year to get something repaired or made. In the mid 1980's, I learned that he had had an out of body experience. I do not know how I learned this. He did not tell me. I had been dealing with him approximately twenty years, and he never mentioned it. After I learned it, I say to myself, the next time he comes in here I am going to ask him about it. Several months go by and one morning he shows up with something to be repaired. After he told me what he wanted done and I had started work on it, I said "Mr. Ott, I understand you had one of those out of body experiences." He said, "Yes, but I don't like to talk about it. People think you are crazy." I could tell from the tone of his voice that I had better let him know right then how I felt about the subject, otherwise he wasn't going to tell me. My answer was that I believed people could have such an experience, and I had read several books by people who had written about their experiences. I also told him I personally knew Dr. George Ritchie, a friend of our family. Dr. Ritchie wrote about his experience in "Beyond Tomorrow." I got to know him and his sister, Mary Jane, as they would spend a week or two at our farm during the summer when they were in their teens. They were the same age as my older brothers and sisters. They lived in Richmond, Virginia, and were next door neighbors to my aunt who visited Cheery Grove about every weekend.

After telling him all this to convince him I didn't think he was crazy, he made no comment. He was a person of few words. There was long silence, probably a minute, and I had begun to think he wasn't going to tell me about it. I think he was processing what he was going to tell me. When he started to speak he didn't stop, but did pause a little bit. He told me it happened when he was a young man in his 20's and got very sick with pleurisy. He was taken to the University hospital in Charlottsville, Virginia which was new then. It would have been what we call state of the art with the most modern equipment at that time. He said it had only been open a year or two. He said they told him they drew two gallons of fluid out of his lungs. It was during this medical procedure in Charlottsville that he had his experience. He would have been near death to have had this much fluid in his lungs as he was not a large person.

After relating the setting and cause , he told of the experience itself. He said, "I saw my moma and my brother. They hadn't been dead very long." I took this to mean from three to five years. He paused, and said "I saw Jesus", and quivered when he said this statement. His body language said way more than his words about seeing Jesus. Another pause, and he said, "It's real pretty there." That's it. That's all he said about his experience. I felt humbled. I didn't ask him any question. I thanked him for what he had shared with me. They say not all communication is with words. One distinct non-verbal message I got from him was, now I have told you this, I don't care if you believe me or not, I know what I saw. It was very real to him. Here he had told me something that happened roughly fifty years back in his life that was still real to him. With the passage of time a lot of details get wiped out, or thought not to be important. This is why I say the Gospel accounts of Jesus life, which were written down more than fifty years after they happened, are not more detailed.

    Now let's look at Mr. Ott's three statements. #one " I saw my moma and my brother. They hadn't been dead very long." These were close family members who he had lived with. I think when we die, we look forward to being united again with close family members. Who dosen"t long to see and hear people we have known, such as parents, brothers, sisters, spouses, children, close friends and be with them again? This is a very strong desire in all of us. He would have recognized them instantly. I could now ask him questions as to how he felt upon seeing them. I was so astounded at the time, I didn't ask him anything.

    # two statement. " I saw Jesus". This is an incredible statement. When I told this to one person, he asked , "how did he know it was Jesus"? Of all the accounts written by people who have had these experiences, I have yet to come across any of them not recognizing Jesus. Jesus is God in human form. The power that comes from Him is so awesome you don't need someone to tell you who He is. For a person to state "I have seen Jesus", with such conviction and not backing down from it , is incredible, and gives us conviction that one day we will meet Him, provided we believe in His existence.

    # three statement. "Its real pretty there." A very simple way of stating the beauty of a place. By using the word 'THERE' this was a very real

A TESTIMONY

place he saw. The book of Revelation, Chapter 21 and part of chapter 22 tries to describe this beauty by the use of precious metals and stones, which we deem beautiful. When I first came across the idea put forth that the Book of Revelations was the apostle John's, out of body experience, I rejected it. After the initial rejection of this idea, I began to study the similarities between what the apostle John wrote and the written accounts of people living today. They all begin by telling of the setting before they have the experience. In Revelations Chapter One V. 9 & 10, John does this. Mr. Ott did this by telling about the illness and being taken to the hospital. Many people tell of being in a horrible accident, where the medical help that first arrives, quickly assess the situation, determining who has vital signs and who does not. The first attention is to those that do have vital signs, and maybe 10 – 15 minutes go by before any attention is given to those that they determined to have none. It is during this time span that these people have these incredible experiences and can state latter what went on while the medical people were working on other people or even them.

In Revelations Chapter 1 V-10, John states "I was in the spirit." What does it mean to be "in the spirit"? His body, just like the people in a car wreck or on a hospital bed are in a spot or place on this earth. Their spirit has left the body and; might be an incredible distance and see incredible things away from the body. In all accounts of people that have these out of body experiences, they use the term "I saw." Everyone of them do this. Mr. Ott did it. Count how many times "I saw" is used in Revelations. This statement is universal to people that tell of their experience. The person having the experience is an observer and remembers what they observed, then comes back and tells people about it. The apostle John was told to write it down, which he did and we have his record. I don't think Mr. Ott did, but he had to have told people in some way what he had seen, and evidently been laughed at, or been called crazy, which had caused him pain. That's why he had the policy of just not talking about it. It is very difficult for our small minds to understand the invisible, non- tangible thing we call "spirit." Some people reject totally the idea of "spirit." The bible says "God is a spirit." John 4 V24. Who are you going to believe? We know for certain that one day our bodies will die. Jesus said, "He who believes in me will never die." John Chapter 11 V-26,

Speaking of our spirit. We have powerful evidence from many people who have had these experiences, that this is true. As Mr. Ott said, you can see again family members and friends who have died, see Jesus and be in a beautiful place.

I trust that Mr. Ott's testimony will help you believe that when your time comes, you too will be united with departed love one's , see Jesus, and be in a beautiful place. A simple message, yet a very profound one.,